The Patrick Moore Practical Astronomy Series

W0079917

More information about this series at http://www.springer.com/series/3192

The NexStar Evolution and SkyPortal User's Guide

James L. Chen
Adam Chen

 Springer

James L. Chen
Shenandoah Astronomical Society
Gore, VA, USA

Adam Chen
Baltimore, MD, USA

ISSN 1431-9756   ISSN 2197-6562   (electronic)
The Patrick Moore Practical Astronomy Series
ISBN 978-3-319-32538-5   ISBN 978-3-319-32539-2   (eBook)
DOI 10.1007/978-3-319-32539-2

Library of Congress Control Number: 2016936674

Cover design by Adam Chen.

Printed on acid-free paper

This Springer imprint is published by Springer Nature
The registered company is Springer International Publishing AG Switzerland

*This book is dedicated to my friends
Helen and Steve,
and to Cheryl and Ray,
for their friendship and support over the years.*

Other Books by
James L. Chen

How to Find the Apollo Landing Sites
(The Patrick Moore's Practical Astronomy Series)
A Guide to Hubble Space Telescope Objects: Their Selection, Location, and
Significance
(The Patrick Moore's Practical Astronomy Series)
The Vixen Star Book User's Guide
(The Patrick Moore's Practical Astronomy Series)

Preface

It is very likely that every amateur astronomer has owned a Celestron telescope at one time or another. And I am no exception. My experiences with Celestron telescopes began at the same time as my passion for backyard astronomy was reborn.

The Washington, D.C., area in the early 1980s was well known for George Washington Birthday sales. Every year, bargain hunters were treated to exceptional deals on all forms of consumer products, such as clothing, furniture, televisions, stereo electronics, and cameras. In 1987, a local Washington, D.C., camera store consolidated all its surplus camera and telescope items into its main warehouse for a true blowout George Washington's birthday sale. This particular sale included several Celestron telescopes that were unsold from the previous year's Halley's Comet sales push. My best friend and I entered the warehouse store and went crazy, with me leaving with two Celestron telescopes and an armful of eyepieces. My prized acquisitions were an orange tube Celestron C-5 with equatorial wedge and an orange tube C-90 Astro with fork mount and clock drive. The Celestron C-5 completely renewed my interest in astronomy. It also sparked a bad case of Gear Acquisition Syndrome, or sometimes known as GAS. In the next decade, I found myself building, buying, and selling many telescopes (including a self-built 10-in. Dobsonian telescope that resulted in my first published article in the November 1989 of Astronomy magazine). Along the way, both orange tube Celestron telescopes were sold. To this day, I wish I had kept that orange C-5.

By the early 1990s, I found myself owning a Celestron Ultima 8. The Ultima 8 was the ultimate expression of a pre-computerized 8-in. Schmidt-Cassegrain telescope, with heavy fork tines and an accurate clock drive system with hand controller. It was heavy, and boy was it stable. It had wonderful optics and was a joy to use, with the exception of having to move it in and out of the house. Alas, apochromatic refractor fever got a hold of me, and the Ultima 8 was sold to finance a Brandon 130-mm apochromat refractor (which I still own). Another Celestron that I wish I had kept.

By the early 2000s, I had bought, traded, and bargained my way through several telescopes, culminating in the ownership of a classic Questar 3-1/2 in. Maksutov-Cassegrain, with a 1/10th wave quartz mirror. It was a wonderfully portable telescope system that accompanied me on a trip to Hawaii, the shores of the Chesapeake Bay to view and photograph the Venus transit of 2004, and several star parties. All was good with the Questar, except for the limitations of such a small aperture. In a clear case of aperture fever, the Questar was traded in for my current big eye telescope, a Celestron 11″ GPS. Eleven inches of aperture, GoTo and GPS drive system, and a versatile 2-in. diagonal, this Celestron has kept me happy for a decade. I don't miss the Questar!

I was working at a vendor booth at the 2014 NorthEast Astronomy Forum, conveniently known as NEAF, and was present at Celestron's product announcement of the Celestron Evolution series of telescopes. At an exclusive Celestron reception, Celestron introduced their new telescope line called the Celestron NexStar Evolution and a new 11-in. Rowe-Ackermann Astrograph.

The Celestron NexStar Evolution represents the latest developments in the long line of Schmidt-Cassegrain designs. The new Celestron Evolution line includes 6-in., 8-in., and 9.25-in. telescopes mounted on newly designed heavy duty single-arm fork mount with WiFi-based computer GOTO drive systems. Mechanically, the Celestron NexStar Evolution newly designed single-arm design is far sturdier than the older SE single-arm design and is steady enough for use in astrophotography.

Most notable is the introduction of a new GOTO computer control system. In the past, all telescope users are familiar with GOTO telescopes, with the hand controller and control cable attached to the base, and the power cables needed to provide power. This rat's nest of cables is eliminated with the new Celestron Evolution telescopes. The telescope base comes equipped with a built-in rechargeable battery. No longer does the user have to lug a separate battery pack to power the telescope or have a power cable cord getting in the way during a observing session.

Additionally, no longer is the telescope user encumbered with an archaic hand controller and the required telephone-like coiled controller cable. The Celestron NexStar Evolution utilizes a revolutionary WiFi interface with the user's tablet or smartphone to control the telescope. The user's iPhone, iPad, or Android tablet or phone is loaded with the SkyPortal application. The SkyPortal application is used to control the Evolution telescope, while providing the useful astronomy information.

The WiFi capability will save you if you have a pet dog like I do. I love my Labrador retriever Kaiser. He's a great dog, but sometimes he gets a little rambunctious and crazy. Not long ago, he got a hold of the Celestron NexStar+ hand control and chewed it up. I found the hand control on the floor, and the connecting cable had been chewed off by Kaiser. What was left of the cable and connectors were found in the corner of the family room, in a pool of yuk. Unfortunately, the hand control bore a slight resemblance to one of his chew toys! He had mistakenly taken the hand control off my computer desk and proceeded to do his dog thing. Hence, I now place all my new hand controls in a glass cabinet for protection. My iPad, with the SkyPortal app, is safe from Kaiser's attacks.

Further NexStar Evolution refinements include tripods that now have gradations imprinted on the extendable legs to aid in leveling the mount on an uneven surface. Of course, a bubble level is built in on the tripod. There are even eyepiece spaces provided in both the tripod and drive base.

With the introduction of the NexStar Evolution series, I realized that a new era had dawned on amateur astronomy, and plans for this book took form. Within these pages, the description and process of using the novel WiFi-based control system provided by the NexStar Evolution and the SkyPortal applications are detailed. Note all photos of SkyPortal in action are taken from the screens of either an Apple iPad or an Apple iPhone 5C. The screens are identical with Android devices.

Clear Skies and Good Music,
James L. Chen

Acknowledgments

A big Thank You to the following people who made this book possible:

To Alan Hale for his invaluable help on the history of Celestron.

To Ed McDonough, Michelle Meskill, Kevin Kawai, Eric Kopit, Bryan Cogdell, and the rest of the Celestron crew for all their technical and historical content support.

To Gary and Sherry Hand of Hands-On-Optics, for providing technical support, conceptual ideas, and encouragement.

To my wife Vickie for her encouragement, her support, and her proofreading and critiquing skills.

To my son Adam for his graphics abilities and valuable photographic suggestions and contributions.

To my son Alex for serving as a soundboard for some of my ideas for the book, and making valuable suggestions and contributions.

And as always,

To Nora Rawn of Springer, who gave a fledgling first-time author a chance, for supporting my book concepts, and being a good audience for my jokes.

Contents

About the Authors

James L. Chen is a Retired Department of the Navy and Federal Aviation Administration Radar and Surveillance Systems engineer. He is a Former Program Manager for Advanced Navigation and Positioning Corporation, guest lecturer at local Washington, D.C./Northern Virginia/Maryland astronomy clubs on amateur astronomy topics of eyepiece design, optical filters, urban and suburban astronomy, and lunar observing, author of an Astronomy Magazine article on Dobsonian telescope design in November 1989 issue, and a contributor to Astronomy Technology Today magazine. His first book was published in June 2014 by Springer, entitled *How to Find the Apollo Landing Sites*. Second book entitled *A Guide to the Hubble Space Telescope Objects* is also available from Springer. Third book entitled *The Vixen Star Book User Guide* is also available from Springer. He served as a part-time technical and sales consultant for two Washington, D.C., area telescope stores for over 30 years.

Adam Chen is a Former Program Manager of media support for NASA Headquarters in Washington, D.C., and creator and executive producer of major NASA publications, including the book and web-book application documenting the history of the Space Shuttle Program "Celebrating 30 Years of the Space Shuttle Program." He served as graphics designer for all three James L. Chen's books and currently works in marketing for Brown Advisory, an investment firm in Baltimore, MD.

Chapter 1

A Brief History
of Computerized
Telescope Mounts
for Amateurs

In the ultimate mating of two hobbies, computers and astronomy, computer controlled telescopes have captured the backyard astronomer's imagination and pocketbook. Known collectively as GoTo telescopes, this advanced technology is fascinating to watch in action as the mount proceeds to point the telescope from object to object with precision, accompanied with the sounds of motors whirring and gears meshing.

A GoTo telescope mount is quite simply a telescope system that is able to find celestial objects in the night sky, and then track them. The GoTo mount can be set up in an alt-azimuth or equatorial fashion, and after the proper alignment procedure, the finderscope is no longer needed for the rest of the evening. Some of the newer GoTo telescopes have electronics and CCD cameras that will perform the alignment procedure automatically.

These telescope mounts are wonderful pieces of technology. The GoTo technology allows for more efficient use of observing time by quickly finding objects in the night sky. Built into the hand controller is a microprocessor, firmware, and built-in memory catalog of the positions of thousands of stars, galaxies, nebulae, open star clusters, globular clusters, planetary nebulae, our solar system planets, and the Moon. Complex algorithms developed and refined over years with improvements in encoders and motor technology have made the GoTo telescope an accepted and desirable telescope feature. Computer controlled telescopes can help it's owner to overcome the fear of looking ridiculous while others watch; no longer will the telescope owner appear incompetent as he tries to find celestial wonders—now he only looks ridiculous as he tries to remember how to set up his telescope!

© Springer International Publishing Switzerland 2016
J.L. Chen, *The NexStar Evolution and SkyPortal User's Guide*,
The Patrick Moore Practical Astronomy Series,
DOI 10.1007/978-3-319-32539-2_1

There is an ongoing debate within the amateur astronomy community on the merits of computer guided and computer controlled telescopes. The hardcore conservative backyard astronomers argue that a beginner or novice individual is better served learning the skies without electronic aids, as generations of stargazers have done. There is merit to this argument. However, in these days of increasing light pollution in urban and suburban neighborhoods, seeing landmark stars used for "starhopping" to locate deep sky objects is becoming increasingly difficult and frustrating to a backyard astronomer, particularly to the beginner or novice. Using bright first magnitude stars for alignment, a computerized GoTo system eliminates frustration and introduces fun into the hobby. The search time for a celestial object is reduced from tens of minutes to mere seconds! With the electronics aiding the observer in finding the deep sky objects, a suburban observer can then take advantage of modern filter technology in overcoming the light pollution in their area. Cheers to the miracle of nebula filters, light pollution filters, and color filters!

Of course, in the worst of urban environments, even using a GoTo telescope and mount can be challenging, especially if bright stars are impossible to see for alignment purposes or otherwise. For instance, in the the middle of brightly lit Las Vegas, the only bright stars visible are Wayne Newton, Celine Dion, and a variety of Elvis impersonators!

The era of computerized GoTo telescopes began in 1984. Computer controlled telescopes took form during the same period as the development of personal computers. During the 1980s, the US telescope company Celestron formed a business relationship with Vixen Company, Ltd of Japan. The American company featured its home grown Schmidt-Cassegrain telescope, while importing the Japanese refractors, eyepieces, and equatorial mounts from Vixen, and marketing them under the Celestron brand. The Sky Sensor was an economical system consisting of a GoTo computer control system with motors designed to attach onto their portable German equatorial mount known as the Super Polaris. The landmark Sky Sensor system was remarkable for its time. As the first consumer affordable GoTo system, it had 472 nebulae, star clusters, and galaxies stored in its memory. This is small, as compared to today's GoTo systems that have 30,000, 40,000, or more deep sky objects stored in their databases.

The reader is cautioned to understand that database claims are sometimes inflated and not necessarily truthful. There are a number of multiple counts for a single object. For instance, the Andromeda Galaxy counts as one object; M31 is an additional object; NGC 224 as another object. Thus the same object is counted as three separate objects in some manufacturer's database claims.

The Sky Sensor was revolutionary in 1984. The Sky Sensor data base contained all the Messier objects, NGC objects brighter than tenth magnitude, and 285 stars brighter than 3.5 magnitude.

Installation of the Sky Sensor onto a Super Polaris mount required a little mechanical dexterity, but could be handled by the end user. And if not, the local dealers were experienced in installing the right ascension motor and electronics card, declination motor and electronics card, gear shafts and pressure plates, and clutch knobs. Plug in the Sky Sensor controller and power supply, and the system was ready for use.

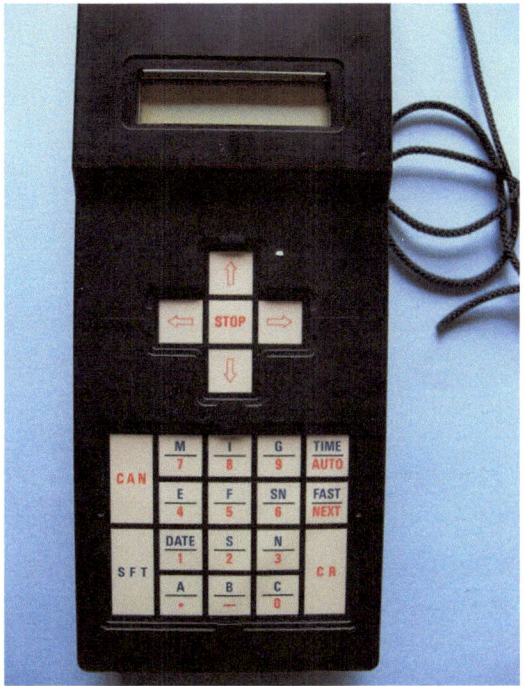

Fig. 1.1 The Sky Sensor computer controller (Hands-on-Optics Used Equipment archives)

The keyboard, as seen in Fig. 1.1 was a bit archaic. Note the use of CR for carriage return instead of an Enter key! The art of human factors engineering had not yet entered into the design of telescope control. The end user faced a bit of a learning curve in operating the Sky Sensor. The system was not as responsive, accurate, nor as quick as today's modern GoTo systems, but as a first generation device it showed the way to the future.

Introduced in 1987, Celestron Compustar 8 was the first computer controlled telescope offered for the consumer. The Compustar 8 was large, heavy, and difficult to produce. The history of Celestron GoTo telescopes is detailed in the next chapter.

In 1992, Meade Instruments introduced the LX200 series of fork mounted Schmidt-Cassegrain telescopes (SCT). Early 8 and 10 in. models that were produced contained software bugs and were unreliable telescopes. Over time Meade was able to refine the LX200 models to become a very capable platform, with the product line extending to larger models, of 12 and 16 in. sizes, telescopes more at home in a college or NASA observatory than in the backyard. In August 1996 Celestron countered with the Ultima 2000 series telescopes—but they delayed shipping until 1997 until the software bugs were worked out. The initial offering was an 8 in. SCT Ultima 2000, which was a lightweight, rigid, and easy to use telescope (Fig. 1.2).

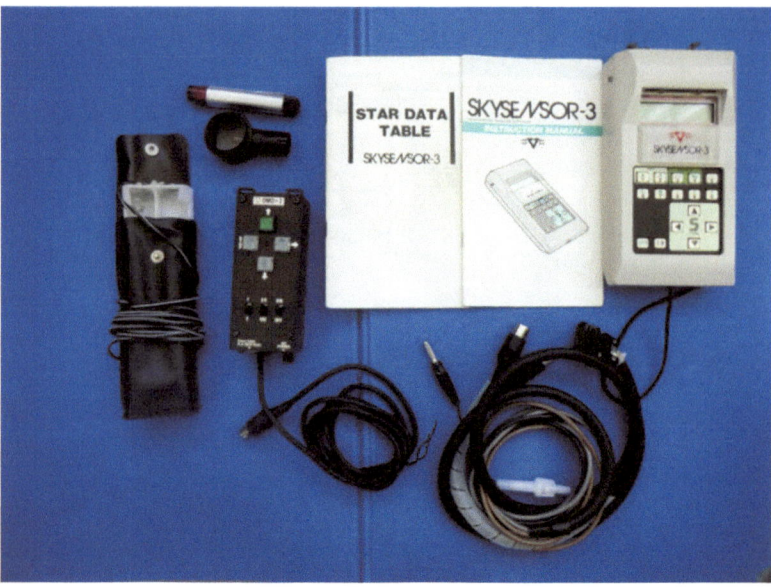

Fig. 1.2 The Vixen Sky Sensor 3 (Hands-on-Optics Used Equipment archives)

Meanwhile in the late 1990s, Vixen issued a revised version of their GoTo system, named the Sky Sensor 3. The Sky Sensor 3 featured an updated hand controller and other hardware. The database was still the same size (Fig. 1.3).

By 2000, Vixen introduced another revision to their venerable Sky Sensor series, now known as the Sky Sensor 2000. The SkySensor 2000 system was vastly refined and improved over the previous Sky Sensor models. The SkySensor 2000 could be retrofitted for use with the Vixen GP, GP-DX, GP-E, SP or SP-DX equatorial mounts to provide highly accurate "Go To" pointing and tracking of celestial objects in a vastly expanded data base that now included the planets, Moon, Sun, and thousands of deep sky objects from Messier, NGC, IC, UGC, SAO, and GCVS catalogs, for a total of 13,942 celestial objects.

The revised system simplified the initial setup and was easier to operate. The slewing rate was improved up to 1200× that of Sidereal rate (5–3/4 deg. per second). The Sky Sensor 2000 incorporated the most accurate of the tracking control systems for the time by including Periodic Error Correction (PEC) circuitry to reduce the amplitude of worm gear periodic errors.

In the new millennia, major developments in GoTo telescope technology have been introduced into the consumer market. Meade and Celestron have introduced and refined their Autostar and Nexstar GoTo systems for fork mount and German mount designs. Databases of these telescope computer systems have been expanded to the 30,000–40,000 celestial objects range, including entire Messier, Caldwell, NCG, and IC catalogs of deep sky objects. Databases now include extensive lists of

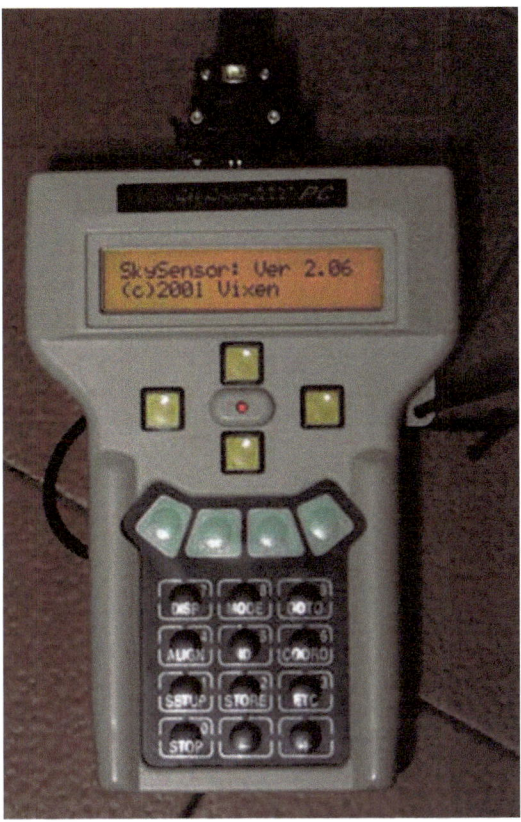

Fig. 1.3 The Sky Sensor 2000 series (Hands-on-Optics Used Equipment archives)

double stars, variable stars, comets, asteroids, and even man-made objects. Pointing precision and tracking accuracies have been greatly improved. The ease of setup has been improved. Many other manufacturers have joined the GoTo mount revolution, with offerings from Losmandy, Orion, Astro-Physics, Takahashi, iOptron, and many more. The computerized GoTo telescope mount has come of age.

Chapter 2

A Review
of Celestron GoTo
Computerized
Telescopes

Celestron is an innovator in the use of computer controlled telescopes, beginning with their introduction of the Compustar 8 and their joint venture participation with the Vixen Super Polaris/Sky Sensor German equatorial mounts.

It is important to be familiar with the various incarnations of Celestron GoTo telescopes and mounts in that SkyPortal with Celestron's SkyPortal WiFi module can be used with many of the legacy Celestron telescopes and mounts.

The best way to chart the historical progress of GoTo telescopes and computerized mounts is to examine the evolution of the Celestron fork mounted GoTo telescopes separately from the GoTo computerized German mounts.

Fork Mounted Celestron SCTs

Celestron Compustar

In 1987, Celestron introduced its line of GoTo computerized Schmidt–Cassegrain telescopes (SCTs) with the introduction of the Compustar line of SCTs. The Compustar system took the DC drive system to the next logical step with a totally computerized go-to telescope. Although the Compustar models were made in 8″, 11″ and 14″ sizes, the only one less than $5000 when new was the Compustar 8 (also sometimes referred to as CCC8). Although somewhat cumbersome with its large declination drive cover and display panel, this unit allowed automatic go-to for over 8000 objects. The f/10 optical tube with "Starbright" coatings featured an 8 × 50 polar axis finder, a 2″ star diagonal and 50 mm 2″ Plossl, heavy duty wedge

© Springer International Publishing Switzerland 2016
J.L. Chen, *The NexStar Evolution and SkyPortal User's Guide*,
The Patrick Moore Practical Astronomy Series,
DOI 10.1007/978-3-319-32539-2_2

Fig. 2.1 Celestron Compustar 8 (Celestron)

and tripod and carrying case. Starting in 1993 the Compustar was shipped with a 1-1/4″ star diagonal and eyepiece with the 2″ accessories as options. The Compustar 8 was the first of the line to be introduced in 1987. It was ahead of its time. Maybe a bit *too* ahead of its time. There have been accuracy and reliability issues. It's mostly a curiosity now, but the model does have some devotees. These innovative telescopes were not cheap for the time, costing $3500 new (Fig. 2.1).

If purchased in the current used market, the buyer should know the Compustar requires a firmware modification to be Y2K compliant. Many of the newer Celestron accessories, such as StarSense, NexRemote, and the SkyPortal Wifi Module, are not compatible with the ancient Compustar electronics.

Celestron Ultima 2000

By 1995, Celestron released a long-expected replacement of its popular computerized telescope. The new model addressed the shortcomings of the original Compustar line. The chief problems with the Compustar were its bulk, weight, and expensive price tag. Celestron addressed all these issues admirably with introduction of the new Ultima 2000, and made the best of new computer tracking technologies as well. The form, fit, and function of the Ultima 2000 is the basis of all subsequent

Celestron GoTo telescopes. Like the competing Meade LX200, the Ultima 2000 shipped without a wedge as part of the standard package. Its computer electronics track objects in the sky after alignment of the telescope on two stars. Due to its built-in high resolution encoders, once the initial alignment is made, the telescope can be turned using the hand controller while still maintaining its position memory. The electronics included Periodic Error Correction. Two motors are used to drive the telescope in each axis, one for slow speeds, and another for high speeds. As a result, the telescope could track normally and be guided at 2× or 6×, or could be moved across the sky at speeds up to 10° per second. The slow and fast slews could be accomplished quietly. The telescope was powered by a set of eight AA batteries mounted in the base, or by an external 12 V power source. The preferred method was external power packs, as the AA batteries were very short-lived. Internal wiring paths for the encoders, etc., meant that the control panel on the base was simplified, but it provided inputs for all the necessary options for a fully operational system. Plugs were provided for external power, electronic focusing, the hand control and an "AUX" port for connecting a computer, an auto-guider or other items. The weight of the optical tube and fork was only 31 lbs. Add another 18 lbs for the sturdy tripod and the Ultima 2000 was a featherweight compared to the old Compustar design. A small hand controller replaced the Compustar control panel. With a few buttons and a two line display, it gave access to all of the aligning, setup, identification, positioning and touring features (Fig. 2.2).

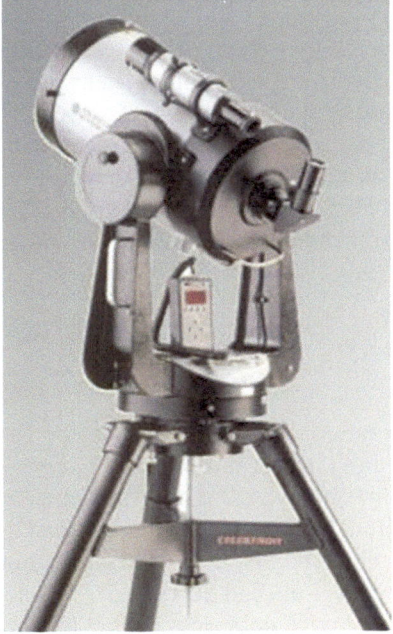

Fig. 2.2 Celestron Ultima 2000 (Celestron)

Fig. 2.3 The author's Celestron 11 NexStar GPS (Chen)

Celestron NexStar GPS

The Ultima series eventually evolved into the NexStar GPS in 2001, with integrated global positioning system circuits to aid in the setup and alignment of the Celestron GoTo telescopes. Again, the aperture sizes ranged from 8″, 9.25″, and 11″. New to this implementation of Celestron SCTs was the introduction of carbon-graphite telescope tubes (although a few aluminum OTAs were available), that functioned to ease the weight of the telescopes while presenting a better thermal adjustment when taken from the inside of a home to the outside environment (Fig. 2.3).

Celestron NexStar SE

The NexStar 8SE, first introduced around 2003 as the NexStar 8i, was the first of the single-sided swing arm ("one armed bandit") computerized scopes. These were given a facelift in the late 2000s with a metallic orange tub, upgraded electronics and the SE nomenclature. The original 2003 version had a gray tube. The newer SE units have the orange tube. The SE line of telescopes ranged from a grab-and-go sized Maksutov–Cassegrain 4 in., and SCTs of 5″, 6″, and 8″. Designed as primarily visual telescopes, their price point was several hundred dollars below the top-of-the-line Celestrons (Fig. 2.4).

Fig. 2.4 Celestron NexStar 8SE (Celestron)

Celestron CPC

In 2005, Celestron introduced its CPC line of computerized GoTo telescopes, programmed with their new SkyAlign firmware. SkyAlign simplified the GoTo initial alignment process by eliminating the "point north and level, then align with specified bright stars" process of earlier designs to a "point at any three bright stars, even if you don't know the name" process. The CPC line included an 8″, 9.25″, and 11″ apertures. The CPC telescopes are characterized by fork mounted SCTs with black tubes. These looked somewhat like the original C8s, but the forks were much thicker and bowed outwards with integrated hand holds for lifting the telescope assembly onto the tripod (Fig. 2.5).

Celestron SLT

In 2001, Celestron introduced a line of intermediate level GoTo telescopes called the GT line, offering 60, 80 and 102 mm refractor models and 114 and 130 mm Newtonian reflector models. In 2005, Celestron introduced a line of intermediate level GoTo telescopes called the SLT line. Utilizing a new upgradeable Nexstar

Fig. 2.5 Celestron 8 CPC (Celestron)

hand control, the NexStar SLT series is quite similar to the Nexstar GT series. In fact, the SLT line of telescopes replaced the GT line, with the exception of those sold through Costco stores (Fig. 2.6).

The NexStar SLT is offered in 60, 80 and 102 mm refractor models and 114 and 130 mm Newtonian reflector models. The two lines are almost identical cosmetically and functionally, with the exception of the hand controller, with the SLT line utilizing the current three star alignment SkyAlign procedure, and the GT line using the previous point-north-and-level procedure.

The differences between the SLT mount and the Nexstar GT are as follows:

1. Tracking performance is greatly improved with the SLT.
2. The GT aluminum tripod has been replaced with a steel leg model on the SLT.
3. The SLT mount includes an AUX port compatible with the CN-16 GPS module and the Auxiliary Port Accessory Kit. The Aux Port Kit allows, among other things, the motor control firmware to be upgraded by the user.
4. The Nexstar SLT is compatible with Celestron's NexRemote software.
5. The optical tube attaches to the fork arm with the ubiquitous Vixen-style dovetail clamp assembly allowing quick and easy removal. In fact, this allows the SLT mount to carry a wide variety of small optical tubes by attaching a matching dovetail bar. The dovetail is compatible with the CG-5, LXD-55, Vixen GPs, etc.
6. The SLT is equipped with a battery compartment for AA batteries for cord-free use.

Fig. 2.6 Celestron 102 SLT (Celestron)

The SLT hand control sports the latest version of the Nexstar firmware. This includes features such as Identify and Constellation Tour. This hand control responds to the same PC commands (via the RS-232 port on the bottom) as the Nexstar 8/9.25/11 GPS, so programs compatible with the Nexstar GPS are compatible with the Nexstar SLT.

Neither the GT nor the SLT featured lock and unlocking clutches on their azimuth or altitude axises, which potentially and practically caused many a telescope to require service for stripped drive gears when youngsters grabbed the telescope and tried to redirect the telescope by hand. All motions had to be directed by the hand control.

The SLT hand control has user upgradeability and the Sky Align three-star alignment methods. Future upgrades to the hand control firmware are downloaded via the Internet and owners can apply the upgrades themselves. The new alignment methods are SkyAlign, Auto Two Star Align (first available on the NexStar 8i Special Edition), Solar System Align and One Star Align. In addition to the new alignment methods, the new hand control still has Two Star Align. Removed since the Nexstar GT are Quick Align and Auto Align.

Fig. 2.7 Celestron 114 LCM (Celestron)

Celestron LCM

At the entry level of Celestron's GoTo telescopes sits the LCM line. Six complete telescope systems with GoTo mounts are available in the LCM series, with apertures of 60, 70, 76, 80, 90 and 114 mm. Designed to be affordable, the telescopes and mount system are light weight and easily portable. The hand controller included in the LCM series, in recognition of the limitations of the optical tubes in the series, is limited to 4000 objects. Still enough to peak a beginner's interest (Fig. 2.7).

SkyProdigy

In a category almost to its own, Celestron has the SkyProdigy line of telescopes, whose main selling point is a fully automated alignment procedure. Mounted parallel to the telescope is a specialized CCD scope used solely for alignment purposes. Mated with complementary on-board firmware, the user merely initiates the alignment process using the hand controller, the telescope begins slewing across the night sky as the CCD scope matches star patterns stored in its memory, and within minutes the telescope is ready for use as a GoTo telescope. This is the initial

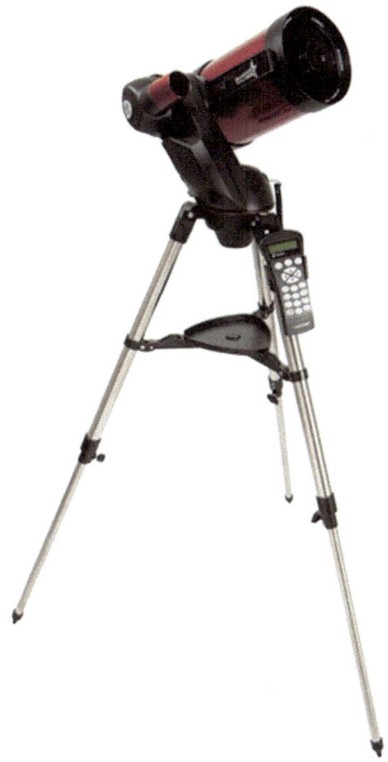

Fig. 2.8 SkyProdigy 6 (Celestron)

implementation of Celestron's proprietary StarSense Technology which provides a completely automatic alignment process with no user intervention required. The GoTo capability contains a database that allows the SkyProdigy telescope to GoTo over 4000 celestial objects. The original lineup of the SkyProdigy telescope included a 102 mm short focus achromatic refractor, a 114 mm Newtonian reflector, a 90 mm Maksutov–Cassegrain, and a 6-in. Schmidt–Cassegrain. Currently, only the 114 mm Newtonian and the 6-in. SCT are being offered (Fig. 2.8).

German Mounted Celestron SCTs

Celestron/Vixen Super Polaris Mount with Sky Sensor

As stated in the previous chapter, the era of computerized GoTo telescopes, and in particular Celestron, began in 1984. Computer controlled telescopes took form during the same period as the development of personal computers. During the 1980s, the US telescope company Celestron formed a business relationship with Vixen Company, Ltd of Japan. The American company featured its home grown Schmidt–Cassegrain telescope, while importing the Japanese refractors, eyepieces, and equatorial mounts from Vixen, and marketing them under the Celestron brand. Prototypes of the Celestron Compustar 14 first surfaced around the 1986–87 time-frame, according to Celestron co-founder Alan Hale. The Compustar 14 is a large and heavy catadioptric telescope, designed for permanent installation in an observatory and not widely available. However, Vixen of Japan developed the Sky Sensor, an economical system consisting of a GoTo computer control system with motors designed to attach onto their portable German equatorial mount known as the Super Polaris. Celestron sold these Sky Sensor-equipped mounts with their 8″ C-8 SCT optical tubes.

The landmark Sky Sensor system was remarkable for its time. As the first consumer affordable GoTo system, it had 472 nebulae, star clusters, and galaxies stored in its memory. This is small, as compared to today's GoTo systems that have 30,000, 40,000, or more stored in their databases (Fig. 2.9).

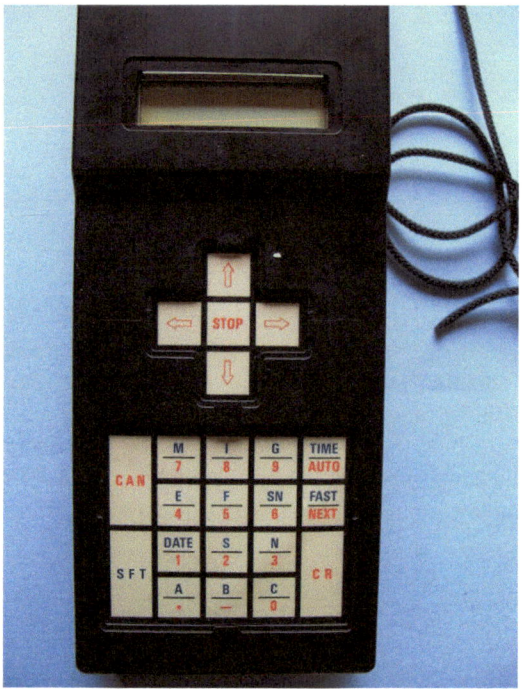

Fig. 2.9 The Sky Sensor controller (Hands-on-Optics photo archive)

Celestron CGE

In the mid-1990s, Celestron introduced a pair of German equatorial mounts for their telescopes for astrophotographers who preferred German mounts over the fork-mounted SCTs. The heavy duty CGE was the top-of-the-line mount, designed to handle the mass of the C-14, but could also accommodate a variety of optical tube assemblies. Fully computerized, the CGE is a very heavy duty equatorial mount with GoTo capability using the Celestron hand controller whose firmware contained a 40,000+ deep space object database. Made from CNC-machined aluminum and stainless steel components, the CGE was designed with a 65-lb. payload capacity, Periodic Error Correction, autoguider port, and other features that made it ideal for astrophotographic applications. For tracking accuracy, Celestron used precision machined bronze 180-tooth worm gear, stainless steel worm shaft, two made-in-USA DC servo drive motors, with dual encoders accurate to 1.5 arc min (Figs. 2.8 and 2.10).

Fig. 2.10 Celestron CGE (Celestron)

The details of the CGE:

1. 40,000 object database with over 100 user-definable objects and expanded information on over 200 objects.
2. Proven NexStar computer control technology.
3. GoTo system is precision accurate to 1.5 arc min.
4. Software Features include: Compass Calibration, Polar Alignment routine, Database Filter Limits, Hibernate, Auto North and Level, Quick Align, and user-defined slew limits.
5. Flash upgradeable hand control software and motor control units for downloading product updates over the Internet.
6. Custom database lists of all the most famous deep-sky objects by name and catalog number; the most beautiful double, triple and quadruple stars; variable star; solar systems; objects and asterisms.
7. Permanent programmable periodic error correction (PEC)—corrects for periodic tracking errors inherent to all worm drives.
8. Precision Bronze Worm Gear—32 pitch, 5.625″ pitch diameter, 180-tooth bronze gear manufactured in the U.S.A. by W. M. Berg, Inc. Manufactured to AGMA Quality Class 10 standard, which minimizes total composite error and backlash.
9. Drive Motors—Made in the U.S.A., Pittman® LO-COGT brush-commutated DC motors offer smooth, quiet operation and long life. The motor armatures are skewed to minimize cogging which is required for low speed tracking.
10. Bearing and Shaft—Stainless steel worm shaft has 0.4375 pitch diameter and is preloaded with two ball bearings. The worm is made from a single piece of steel (rather than a two-piece worm-and-shaft assembly) to minimize run-out, which is a source of periodic error.
11. 12VDC Servo Motors with integrated optical encoders with 0.11 arc sec resolution.
12. No-slip clutch system for pointing precision.
13. Autoguider port, PC port and auxiliary ports located on the electronic pier for long exposure astrophotography.
14. Double line, 16-character Liquid Crystal Display Hand Control with backlit LED buttons for easy operation of goto features.
15. RS-232 communication port on hand control to control the telescope via a personal computer.
16. Includes NexRemote telescope control software, for advanced control of your telescope via computer.
17. GPS-compatible with optional CN16 GPS Accessory.

Celestron AS-GT (CG-5GT)

The second of the mid-1990s GoTo mounts was the computerized version of the venerable CG-5, anointed as the AS-GT, but commonly referred to as the CG-5GT. These GoTo versions of the CG-5 also come loaded with new software

Fig. 2.11 Celestron AS-GT (Celestron)

features and over 40,000 database objects. Capable of holding over 35 lbs of pay-load and slewing at 4° per second, users are able to instantly point to any of the celestial objects in the database (Fig. 2.11).

The Advanced GT Series telescopes include these high performance features:

1. GPS-compatible with optional SkySync GPS Accessory (#93969).
2. Proven NexStar computer control technology.
3. RS-232 communication port on hand control to control the telescope via a personal computer
4. Autoguider port for long exposure astrophotography.
5. 40,000+ object database with 100 user-definable objects and expanded information on over 200 objects.
6. Custom database lists of all the most famous deep-sky objects by name and catalog number; the most beautiful double, triple and quadruple stars; variable star; solar system objects and asterisms.
7. Double line, 16-character Liquid Crystal Display Hand Control with 19 fiber optic backlit LED buttons.
8. DC Servo motors with encoders on both axes.
9. Affordable price.

Celestron CGEM, CGEM DX

Celestron's CGEM mount fits between the Advanced Series and CGE Series. Offering the portability of the Advanced Series and the precision of the CGE and is capable of carrying Celestron's higher-end SCT optical tubes (up to 11″) securely and vibration free, which is ideal for both imaging and visual observing. The mount is capable of holding over 40 lbs of payload and slewing at 5° per second (Fig. 2.12).

The CGEM was designed to be ergonomically friendly with large Altitude and Azimuth adjustment knobs for quick and easy polar alignment adjustment. The internal RA and DEC motor wiring provides a clean look and an easy and trouble free set up.

The CGEM series introduced a new innovative Polar alignment procedure called All-Star™. All-Star is a relative of the AltAz SkyAlign, in that it allows users to choose any bright star, while the software calculates and assists with polar alignment. Another feature of the CGEM available for astro-imagers, is the Permanent

Fig. 2.12 Celestron CGEM (Celestron)

Periodic Error Correction (PEC) which will allow users to train out the worm gears periodic errors, while the mount retains the PEC recordings.

For objects near the Meridian (imaginary line passing from North to South), the CGEM will track well past the Meridian for uninterrupted imaging through the most ideal part of the sky. The CGEM mount has a robust database with over 40,000 objects, 100 user defined programmable objects and enhanced information on over 200 objects.

Celestron CGE Pro

In January, 2009 Celestron introduced the heavy duty CGE Pro model. CGE Pro is a step up on capacity and stability from the original CGE series. Celestron CGE Pro series is a heavy duty German equatorial mount (GEM) on a steel-legged tripod, utilizing a German equatorial mount specific NexStar hand controller with Sky Align. The series includes 9.25, 11 and 14 in. SCT models. The CGE Pro can also be purchased sans telescope. Think of the CGE Pro as a CGE on steroids (Fig. 2.13).

Fig. 2.13 Celestron CGE Pro (Celestron)

The CGE Pro series is a very stable platform for a wide variety of optical tubes. It is an excellent platform for both visual and photographic work and is an excellent choice for a permanent observatory setup. For those looking for a more stable astrophotography platform, particularly for imaging with a larger SCT optical tube with many accessories, the CGE Pro is definitely the choice over the CGE and CGEM mounts.

The Celestron CGE Pro features are:

1. Holds a maximum instrument capacity of 90 lbs.
2. Smooth ±5 arc sec typical unguided periodic error, which can be further reduced with PPEC.
3. Permanent Programmable Periodic Error Correction (PEC)—corrects for periodic tracking errors inherent to all worm drives.
4. Heavy duty stainless steel tripod adjustable from 38″ to 55″.
5. One 22 lbs counterweight.
6. Weighs 154 lbs.

In addition to being fully computerized with a database of over 40,000 celestial objects, the New CGE Pro German Equatorial mount has been completely redesigned to offer numerous design advantages for the Astrophotographer:

1. Increased Load Capacity—Able to hold our 14″ SCT telescope more securely as well as larger optical tubes up to a maximum instrument capacity of 90 lbs (not including counterweights).
2. All-Star Polar Alignment—Choose any bright alignment star for a software assisted alignment of the mount's polar axis that will have the user ready for imaging even if the North Star can't be seen.
3. No Tool Polar Alignment—Larger hand knobs for both Altitude and Azimuth adjustments.
4. Meridian tracking—Extended tracking past the Meridian of up to 20° of uninterrupted imaging through the best part of the sky.
5. Faster Slew Speed—Improved gearing and motors provide faster slew speeds than ever before with a maximum slew rate of over 5° per second.
6. Power Management—Redesigned electronics deliver constant regulated power to the motors making them capable of driving the telescope even when not perfectly balanced. This allows the CGE Pro to have the payload capacity of that of much larger (and expensive) mounts without sacrificing smooth tracking motion and pointing accuracy across the entire sky.
7. Accuracy—The hallmark of any telescope mount is its ability to find, center and track celestial objects with the highest degree of accuracy. The CGE Pro provides the precision pointing and tracking accuracy needed for demanding visual and imaging pursuits.
8. Pointing—With just a standard hand control alignment, the CGE Pro has the ability to center a star in the eyepiece or CCD chip to within 5 arc min. Using NexStar's advanced pointing features such as Calibration Stars, Sync and Precise GoTo, further improves the pointing accuracy to as low as 1 arc min in the desired region of the sky.

9. Tracking—With larger .75″ pitch diameter worm gear and 6″ pitch diameter worm wheel, precision made cut-steel gears in gearboxes, and seven slot-skewed armature motors, the CGE Pro delivers smooth ±5 arc sec tracking accuracy typical unguided periodic error, which can be further reduced with PPEC.

10. Mount calibration—Celestron's NexStar hand control has built-in compensation features essential for accurately placing small objects on the center of the CCD chip or high power eyepiece. Aligning on multiple Calibration Stars creates a model of the mechanical inaccuracies inherent in all equatorial mounts. This model is stored within the hand control and is used to compensate for these inaccuracies, thus improving the pointing precision each time the telescope is slewed.

11. Mount modeling—Celestron's NexStar hand control has built-in mount modeling features essential for accurately placing small objects on the center of the CCD chip or high power eyepiece. Aligning on multiple Calibration Stars creates a mathematical model of the mechanical inaccuracies inherent in all equatorial mounts. This model is stored within the hand control and is used to compensate for these inaccuracies, thus improving pointing precision each time the telescope is slewed.

12. And of course, the CGE Pro is also fully T-Point compatible (available from Software Bisque) for ultra precise pointing across the entire sky.

In addition to these improvements, the Celestron line of German Equatorial mounts has long been recognized for features preferred by visual observers and astrophotographer alike. These include:

1. Portability: Set up and transportation of the CGE Pro telescopes is made easy by separating the mount into smaller, easy-to-carry components. Unlike fork arm mounted telescopes, the CGE's optical tubes can be quickly removed from their mounts making even the CGE Pro1400 easily assembled in minutes.

2. Stability: Recognized for superior stability, German Equatorial mounts place the center of gravity directly over the tripod legs and can be easily polar aligned without the use of an optional equatorial wedge. This proven design reduces the "tuning fork" vibration that can be associated with undersized fork mounts. An improved Super HD Tripod supports the CGE Pro mount. This fully extendable tripod is made from the finest 2.75″ stainless steel and can be raised to a height of 55″. The tripod uses a dual leg support for maximum rigidity with an upper leg brace to provide an outward preload and a lower leg brace providing inward tension.

3. Balance: CGE Pro Equatorial mounts can easily be balanced in both axes. Simply sliding the counterweight for Right Ascension and moving the optical tube along its dovetail mounting for Declination, accomplishes balancing the weight of camera equipment and other visual accessories. This means that no additional weight needs to be added to balance the telescope when additional accessories are added.

4. Clearance: CGE Pro mounts support their tubes at a single contact point allowing the tube to move freely around its polar axis without making contact with the telescope's mount. Software features allow the user to set the mounts slew limits to guaranty safe motion. This is particularly useful when adding photographic and CCD instruments that extend from the rear of the telescopes.
5. All CGE mounted telescopes are compatible with Celestron's SkySync GPS accessory. Combine the GPS and built-in real time clock, and these telescopes will keep track and remember their exact location and time without having to enter the information into the hand control.

Celestron VX

Introduced shortly after the CGEM and the CGE PRO, the VX is the follow-on replacement model for the venerable CG-5GT (Fig. 2.14). The essential features of the VX are:

1. Holds a maximum instrument capacity of 30 lbs.
2. Integer gear ratios and permanently programmable Periodic Error Correction eliminates recurring track errors from the worm gear.
3. New motors offer improved tracking performance and provide more power to overcome load imbalances.
4. Updated industrial design offers more rigidity, less flexure and improved aesthetics.
5. New design allows viewing or imaging across the meridian without interference from the motors housings.
6. Improved latitude range. Can be used between 7° and 77° latitude.
7. Improved electronics with increased memory for future expansion.
8. NexStar+hand control offers multiple language programming (English, French, Italian, German, Spanish).

Engineered from the ground up with astro-imaging in mind, the new Advanced VX series improved on the standard set by the CG-5GT in mid-level telescopes. Advanced VX provides the user with many of the features found on Celestron's most sophisticated German equatorial mounts, at an extremely affordable price.

The new Advanced VX mount was specifically designed to provide optimum imaging performance for smaller telescopes. With the Advanced VX, owners of smaller telescopes can take advantage capabilities that the CGEM and CGE Pro users have been enjoying, such as All-Star Polar Alignment and autoguider support. Tracking through long exposures using permanently programmable periodic error correction is available. The amateur astro-imager can now image across the meridian without doing a meridian flip, so the backyard astronomer can seamlessly image the best part of night sky. Advanced VX features significantly larger base castings than the CG-5GT design it replaces, improving stability under heavier loads. Improved motors offer more torque and can handle slight load imbalances with ease.

Fig. 2.14 Celestron VX (Celestron)

Chapter 3

Introduction to the Celestron NexStar Evolution and SkyPortal App

The annual springtime ritual that is the North East Astronomy Forum, commonly known as NEAF, brought an added level of excitement in 2014. Celestron used NEAF 2014 as an opportunity to introduce new products to their extensive arsenal of telescope offerings.

At an exclusive Celestron reception, Celestron introduced their new telescope line called the Celestron NexStar Evolution, along with a new 11-in. Rowe-Ackermann Astrograph.

The Celestron NexStar Evolution represents the latest developments in the long line of Schmidt-Cassegrain designs. The new Celestron NexStar Evolution line includes a 6-in., 8-in., and 9.25-in. telescopes mounted on a newly designed heavy duty single-arm fork mount with wifi-based computer GoTo drive systems.

Mechanically, the Evolution series features a newly designed single-arm fork mount that is far sturdier than the older SE design, and is sturdy enough for use in astrophotography.

Most notable is the new GoTo computer control system. In the past, GoTo telescopes had hand controls connected by a control cable attached to the base, and a power cable attached to the battery or power supply. This rats nest of cables have been eliminated in the new Evolution telescopes.

No longer is the telescope user encumbered with a hand controller and the required telephone-like coiled controller cable. The Celestron NexStar Evolution utilizes a cordless revolutionary WiFi interface with the user's tablet or smartphone to control the telescope. The user's iPhone, iPad, or Android tablet or phone loaded with the SkyPortal application, a special version of the Sky Safari 4 application, is used to control the Evolution telescope, while still providing all the useful

© Springer International Publishing Switzerland 2016 27
J.L. Chen, *The NexStar Evolution and SkyPortal User's Guide*,
The Patrick Moore Practical Astronomy Series,
DOI 10.1007/978-3-319-32539-2_3

astronomy information that made Sky Safari famous. A standard Nexstar+ hand controller is included with each Evolution telescope to provide an alternative means of controlling the telescope.

Additionally, the telescope base comes equipped with a built-in rechargeable lithium iron phosphate battery battery. This battery, as known as LiFePO4, includes numerous built-in safety features including protection against overcharging, over-discharging, and overheating. Lithium iron phosphate batteries are long-life units that under normal usage will last through thousands of charge cycles before replacement is needed. No longer does the user have to lug a separate battery pack to power the telescope, or have a power cable cord getting in the way during a observing session.

Lithium iron phosphate batteries offer a longer cycle life and longer calendar life relative to the more traditional battery chemistries, with the following advantages:

1. LiFePO4 batteries have higher current and peak-power ratings than LiCoO2 batteries.
2. The use of phosphates avoids certain environmental concerns, particularly about cobalt entering the environment through improper disposal.
3. High charge levels and elevated temperatures (whether from charging or ambient air) hasten capacity loss of LiCoO2 cells. In contrast, the calendar life of LiFePO4 cells is not affected by high charge states.
4. Unlike other lithium ion batteries, LiFePO4 can deliver virtually full power until it is discharged.
5. One important advantage of LiFePO4 over other lithium-ion chemistries is thermal and chemical stability, which improves battery safety.

The NexStar Evolution telescopes are equipped with 4 AUX ports for attaching the NexStar+ hand control and other Celestron accessories. Also included is a USB power output port, available for charging iOS and Android devices.

Further refinements are included in the tripods that now have gradations imprinted on the extendable legs to aid in leveling the mount on an uneven surface. Of course, a bubble-level is built-in on the tripod. There are even eyepiece spaces provided in both the tripod and drive base (Figs. 3.1, 3.2, and 3.3).

Telescope Optics

Catadioptric telescopes are a category of telescopes that combine lens and mirror technology to produce compact and transportable instruments. With a clever combination of a lens and mirrors, the incoming light path is folded upon itself, and any optical aberrations of the reflecting surfaces can be corrected by the refracting lens (Fig. 3.4).

Catadioptrics are available in two popular forms: Schmidt-Cassegrain telescopes, or SCT and the Maksutov-Cassegrain telescope. Maksutov-Cassegrains appeared on the commercial market in the mid-1950s, and SCT's burst into

Fig. 3.1 Celestron NexStar Evolution 8 (Celestron)

popularity in the 1970s. As a family, catadioptrics offer a high degree of optical performance and low maintenance in a small compact package.

All three Evolution telescopes are of the Schmidt-Cassegrain design, which was first brought forth to the amateur astronomy market and popularized by Celestron in the mid-1960s. The three sizes of Evolution telescope represent realistic choices of convenience, optical performance, and portability.

Although SCTs are available in sizes from 5-in. to massive 14 and 16 in. apertures, the most popular and best-selling telescope since the 1970s has been the 8-in. SCT. The 8-in. SCT is the basic core product for not one, but two major telescope companies. The reason for this popularity stems from the all-around versatility that the design provides. As an analogy, the 8-in. SCT is the Olympic Decathlon Champion of telescopes. The Olympic Decathlon gold medalist doesn't excel in any singular competition, such as the 100 m dash or the high jump, but is able to perform ten different events better than anyone else can perform those ten same events. Versatility and a high level of performance are outstanding attributes of an 8-in. SCT. The 8-in. aperture allows for light gathering for deep sky objects, the typical f/10 focal length enables high magnification for lunar and planetary

Fig. 3.2 Celestron NexStar Evolution 6 (Celestron)

observations, numerous attachments are available for photographic use, and the compact size ensures portability and frequent usage. But it's not perfect for any singular use. Newtonian telescopes are available at a cheaper price in larger sizes for deep sky light gathering. Refractors produce images that are of better quality in contrast and sharpness. Many amateur astronomers tend to own two, three, or more telescopes in order to optimize their viewing. But if a person is only going to own one single telescope that is capable of handling multiple astronomy tasks, the Evolution 8-in. SCT is the likely choice.

However, the 9.25-in. optical tube is widely regarded as something uniquely special in the Celestron's SCT model range. Its primary mirror has a larger focal ratio, f/2.5 rather than f/2, than other models, so the secondary magnification factor is less, resulting in a smaller secondary obstruction. The smaller secondary obstruction allows for better contrast and a flatter field which is a distinct advantage for imaging. All things considered, if funds allow, the best choice for maximum performance is the Evolution 9.25.

For maximum convenience and portability, combined with superb optical performance, the Evolution 6 is the choice. Its convenient size and still ample aperture fills the need for a grab-and-go telescope for trips to the country at a moments notice. Much of the research for this book was based on a NexStar Evolution 6.

Fig. 3.3 Celestron NexStar Evolution 9.25 (Celestron)

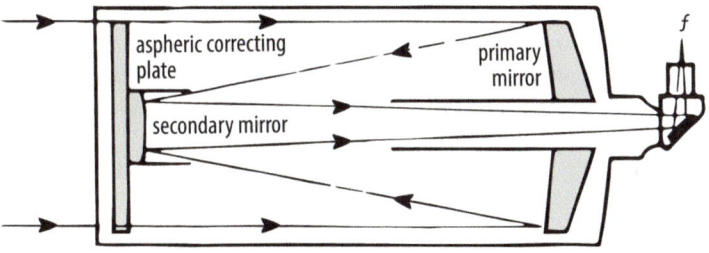

Fig. 3.4 Schmidt-Cassegrain design (Adam Chen)

SkyPortal Application

At the heart of the Celestron WiFi technology is the SkyPortal app. Celestron's newest planetarium app is an astronomy suite, based on Sky Safari 4 Basic, that redefines the computerized telescope experience of the night sky. SkyPortal enables the backyard astronomer to explore the Solar System, 120,000 stars, 220 star clusters, nebulae, galaxies, and dozens of asteroids, comets, and satellites—including the ISS.

SkyPortal is compatible with both Apple iOS and Android devices. The user has the option of using a smartphone, a mini-tablet, or full-sized tablet. SkyPortal is free for download and is available from the online Apple App store and Google Play.

Additionally, SkyPortal and the smart device can be used with either the compatible Celestron NexStar Evolution telescope, or with any GoTo Celestron mount with the added SkyPortal WiFi module.

As of this writing, the current version is SkyPortal Version 1.5.17, supports the following:

1. Support for Alt Az telescopes using an equatorial wedge.
2. Improvements to night vision mode.
3. Improvements to the user interface for using saved settings.
4. Added support for push notifications.
5. Fixed problem where some planet surfaces and horizons would not draw correctly during an update of the orbital elements.
6. Stability improvements.
7. Compatible with the Celestron StarSense Auto Align accessory.

SkyPortal Planetarium Features are as follows:

1. SkyPortal's intuitive Compass Mode enables the user to hold the smart device up to the night sky and instantly identify stars, planets, galaxies, and more. Zoom in to view fainter objects not visible to the naked eye.
2. View a custom list of all the best celestial objects to view based on the local time, date, and location.
3. Simulate the night sky up to 100 years in the past or future to plan a particular observing session. For example, the user can look ahead to see when Jupiter's Great Red Spot will be visible. Or animate a lunar eclipse such as the "Blood Moon" to prepare the user on what to expect before setting up the telescope.
4. View hundreds of astronomical photographs and NASA spacecraft images. Or listen to more than 4 h of audio narration to learn the history, mythology, and science of the heavens.
5. Night Vision Mode helps you preserve your eyesight after dark.

Upgrades to SkyPortal

For many owners of an Evolution telescope, the capabilities of SkyPortal is sufficient to keep them happy for years. 120,000 stars, plus 220 of the best-known star clusters, nebulae, and galaxies in the sky are available with SkyPortal, with SkySafari 4 serving as the development foundation. It displays the Solar System's major planets and moons using NASA spacecraft imagery, and includes the best-known 500 (or so) asteroids, comets, and satellites. It accurately shows the sky from anywhere on Earth, at any time up to 100 years in the past or future, and lets the user identify stars, planets, and constellations with the smart device's GPS, compass and/or gyroscope.

But many owners may feel limited by the capabilities of SkyPortal, especially considering the extensive deep sky capabilities of the NexStar+ hand control provided as standard equipment with each Evolution telescope. The Celestron NexStar+ AZ Hand Control is a standard accessory on all current altazimuth mounted NexStar telescopes, and is compatible with the Celestron SLT, LCM, SE, and CPC computerized telescopes.

With the push of a button, the user can access NexStar+ hand control's huge database of over 40,000 celestial objects, automatically slewing to objects from a variety of catalogs, including the Messier, NGC, Caldwell, and SAO brightest stars. The planets as well as a list of the most popular objects are also included.

Not sure what to look at? Sky Tour provides a selection of the most popular objects for the date and time of night.

A variety of alignment procedures are pre-programmed into the NexStar+ Hand Control, including SkyAlign, Auto 2-Star, 1-Star, 2-Star, and Solar System. A variety of tracking rates (sidereal, solar and lunar) and slew speeds are available. Slew rates can be chosen at 1°, 2°, or 3° per second, or choose speeds as low as 2× or as high as 64×.

This hand control includes Flash Upgradeable Technology, allowing users to upgrade its software via the Internet. Access the Celestron website and the latest upgrades can be found on the Support Tab.

One of the benefits of SkyPortal versus the NexStar+ hand control is the method of upgrading the software. The NexStar+ hand control requires a connection to a PC computer and adjustments to port assignments and addresses. In other words, a little computer knowledge and technical skill that many people lack. There is a risk of frustration and failure with the process of upgrading a hand control. Updates with smart devices is a simple download of the latest app version. The user is assured prior to the download that the app will work properly and install properly because of the testing and vetting process required prior to its availability through the app store. There is little or no risk associated with upgrading an app on a smart device.

For those who find the databases of SkyPortal (which is based on SkySafari 4 Basic) and the NexStar+ hand controller limiting, SkySafari 4 Plus and SkySafari 4 Pro are compatible apps for the smart devices and the Evolution telescopes. Both can be purchased from either Apple App Store or Google Play.

SkySafari 4 Plus adds a hugely expanded database, wired or wireless telescope control, and the ability to leave Earth and fly into orbit around any Solar System object or nearby star—from the SkyPortal basic version. It shows you 2.6 million stars, and 31,000 deep sky objects—including the entire NGC/IC catalog. It includes nearly 18,000 asteroids, comets, and satellites with orbits that can be updated. And it can point the Evolution telescope anywhere in the sky, using the smart devices built-in WiFi. Many of the additional objects will challenge detection with the apertures of the Evolution series.

The all-new SkySafari 4 Pro has the largest database of any astronomy app, period. It contains everything in SkySafari 4 Plus—but also includes over 27 million stars from the Hubble Guide Star catalog generation 1 and 2, plus 740,000 galaxies down to 18th magnitude, over 620,000 solar system objects—including every comet and asteroid ever discovered—and a Moon map based on NASA's latest LRO data with 8× the resolution of any other SkySafari version. It shows the sky with sub-arc second precision from anywhere on Earth, in the Solar System, or beyond, at any time up to one million years in the past or future—yet it runs just as fast and smoothly as SkyPortal. The caveat to the use of SkySafari Pro is many of the additional objects may not be within the visible or detectable capabilities of the 6″, 8″, or 9.25″ Evolution telescopes.

Chapter 4

Basic Operation of the Celestron NexStar Evolution and SkyPortal App

The operation of the Celestron NexStar Evolution and its SkyPortal app is closely tied with the initial setup of the mount, the WiFi network, and the smart device hosting the SkyPortal app. Mistakes and errors during the initial setup are the major contributors in owners encountering problems with any computerized GoTo telescope/mount system, and the Celestron equipment and SkyPortal app is no different.

The physical setup of the Celestron NexStar Evolution is straight forward, and well-covered in the Celestron NexStar Evolution owner's manual.

Evolution Physical Setup

Assembly of the tripod is simple and intuitive, with the accessory tray being placed over the central column and the tripod support nut and washer threaded onto the threaded column and tightened. The tripod support nut will work either way, so if the threads of column don't seem to tighten one way, feel free to flip the nut over and try again. It will work and tighten.

Placing and securing the Evolution to the tripod is merely the tried-and-true centering the mount over the center pin on the tripod head. Once the centering pin is securely in the mounting hole in the base of the mount, rotate the 6″ or 8″ Evolution mount until its settles, or clicks, into the detents of the tripod head and the mounting sockets line up with the captured bolts. The 9.25″ Evolution does not click into place, as noted in the manual. Thread the captured mounting bolts from underneath the tripod, and tighten them securely.

© Springer International Publishing Switzerland 2016
J.L. Chen, *The NexStar Evolution and SkyPortal User's Guide*,
The Patrick Moore Practical Astronomy Series,
DOI 10.1007/978-3-319-32539-2_4

The Evolution 6 is delivered with its 6″ OTA already attached to the mount. The Evolution 8 and 9.25 OTA are shipped separately from their single arm Evolution mount. The OTA's are attached to the mount via a universal dovetail bar attached to the OTA that fits into the dovetail mount bracket on the Evolution mount arm. The quick-release knob is tightened to secure the OTA onto the mount. Prior to use, loosen the quick-release knob slightly and move the OTA forward or back upon the telescope with accessories, such as the diagonal, eyepiece, and red-dot finder attached, is balanced and re-tighten the knob to secure the OTA.

A mistake that often occurs in the setup and use of GoTo telescopes is the failure of novice backyard astronomers to properly align their red dot finder, or finder-scope if so equipped. The Celestron manual promotes the use of a bright planet or star to align the StarPointer red dot finder, this really is not necessary. Rather than struggling in the dark, pick a time late in the day at dusk to perform this procedure. The target and the red dot will both be readily visible. Pick out a distant stationary target that is at least 1 mile away, such as a stop sign, a light on top of a cellphone antenna tower, or a neighbor's house light. This works better than a planet or bright star that moves as the Earth turns. Adjust the StarPointer red-dot finder as per the instructions.

The physical setup of the Evolution is done.

Smart Device and WiFi Setup

Setting up the wifi connection between the Evolution and the smart device is fairly simple, requiring the user to be familiar and comfortable with entering the their smart device's settings menu and making the necessary settings. As Celestron recommends, practicing the procedure during the daytime prior to the initial night use is advantageous.

The wifi connection procedure is as follows:

1. Turn the Evolution switch on. Observe that the Celestron logo LED will illuminate, along with the WiFi LED blinking, while the system awaits a wifi connection.
2. Go to the smart phone or tablet Settings menu and access the wifi settings. The smart device will identify all the wifi networks within the device's range. Select the network that is identified as "SkyQLink-xx". The device will connect to the telescope. (Note: the term SkyQLink is the legacy Celestron term for the wifi link, and was used by Celestron in their early wifi modules. The current SkyPortal WiFi module replaced the prior models SkyQLink and SkyQLink2 modules. The use of these modules are discussed in a later chapter.) While in the smart device's settings, disable the Sleep or Auto-Lock mode of the smartphone or tablet. More on this in a later chapter.
3. On the smart device, launch the SkyPortal app. If it doesn't automatically start on the "Connect and Align—Connect" screen, simply tap the telescope icon to bring up the appropriate screen.

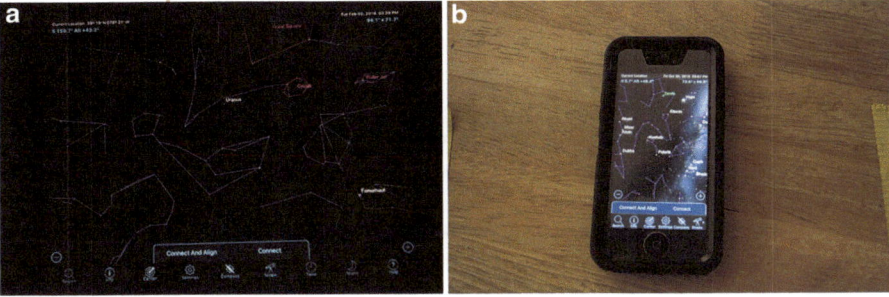

Fig. 4.1 (**a**) The connect and align and connect screen—iPad Tablet(Chen) (**b**) The connect and align screen—iPhone (Chen)

4. If this is the initial setup for the observation session, tap the "Connect and Align" and the SkyPortal will take control of the smart device and telescope direct connect wifi network, and then enter SkyAlign.
5. If reconnecting the SkyPortal app, tap the "Connect" to resume SkyPortal control of the smart device and telescope wifi network (Fig. 4.1a, b).

Note: Figure 4.1b represents the SkyPortal screen on an iPhone. Similar view will be seen when using an Android phone. For purposes of clarity and scale, iPad tablet photos of the screen will be used henceforth in this book.

SkyPortal, Alignment Options and SkyAlign

The basis of all GoTo telescope operations is aligning the telescope, which is the process of telling the computer and software the time, site location, and positions of bright reference stars in the sky. In many GoTo systems, the user must input the current local time and latitude and longitude position. In some cases, the telescope must be pre-positioned prior to the alignment in a "Home" position. Then the process proceeds to finding and centering two, three, or more bright known stars in the finder and telescope eyepiece.

The SkyPortal app and SkyAlign simplifies the alignment process. Since SkyPortal is resident on an iOS or Android device, the time and GPS coordinates are automatically available for SkyPortal to use. Under the SkyPortal Settings, three methods of aligning the telescope and mount are available:

1. Align using SkyAlign—This is the default and best alignment process for the Evolution telescope. When Connect and Align is selected, the Evolution wifi connects to the smart device and SkyPortal stands ready for alignment by centering and aligning with three bright stars in the telescope's eyepiece. Each star is then centered using the SkyPortal directional arrows. The user need not know the identity of the alignment star.

2. Align using Manual Align—This is a process similar to SkyAlign in the use of three bright stars to align, but the user chooses knows the name of the stars. The star is selected on the smart device screen. The star is then centered using the SkyPortal directional arrows.
3. Align using Wedge Align—This option is used when the optional equatorial wedge is used to mount the Evolution, or when using the optional SkyPortal WiFi module on an appropriate Celestron German equatorial mount. The actual alignment process is similar to the Align is Manual Align. The SkyAlign process does not work with an equatorially mounted telescope.

Prior to aligning the telescope, place the Celestron provided standard equipment 40 mm eyepiece into diagonal of the telescope. Advanced amateur astronomers may own a low power and wider field eyepiece and this can also be used. Make sure the red dot finder is properly aligned.

The tapping of the "Connect and Align" will connect SkyPortal to the WiFi network, and proceed through the SkyAlign procedure. Figure 4.2 will appear on the screen, noting that "0 of 3 Stars Aligned", and the user is prompted to:

1. Use the directional arrows on either side of the screen to direct the Evolution telescope towards a bright star. The user is not required to know the name of the star, SkyAlign will figure that out.
2. Once the bright star is found by placing the red dot on the target star and the star is within the eyepiece field of the telescope, tap "Enter".

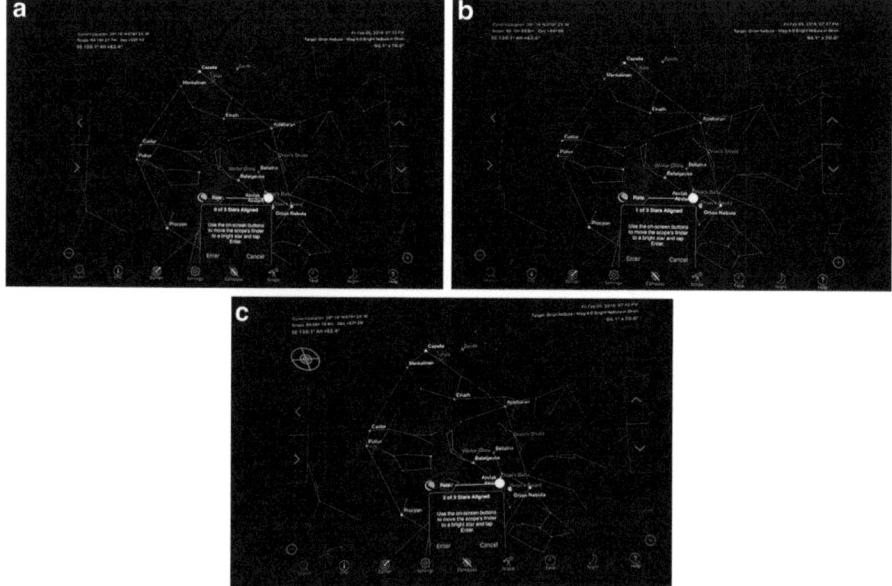

Fig. 4.2 (a) First alignment screen for SkyAlign (Chen) (b) Second alignment screen for SkyAlign (Chen) (c) Third alignment screen for SkyAlign (Chen)

3. Using the directional arrows, center the target alignment star in the center of the eyepiece field.
4. Once the target alignment star has been centered in the eyepiece field, tap align. The SkyPortal screen will display "1 of 3 Stars Aligned".
5. Repeat Steps 1 though 3 on a second target alignment star.
6. Once the second target alignment star has been centered in the eyepiece field, tap align. The SkyPortal screen will display "2 of 3 Stars Aligned".
7. Repeat Steps 1 though 3 on a second target alignment star.
8. Once the third target alignment star has been centered in the eyepiece field, tap align. The SkyPortal screen will display "Alignment Successful".
9. The telescope is now aligned, and the observing session can now begin.

If something has gone wrong, and this screen appears (Fig. 4.3):
A "Alignment Failed" message is an indication of any number of problems:

1. One or more of the target alignment stars was improperly centered.
2. One or more of the target alignment stars was not a first or second magnitude star, and SkyAlign did not recognize it as a possible alignment star.
3. Sometime during the alignment procedure, the user has accidentally bumped the telescope or tripod leg and changed the physical position of the telescope. This happens more often than one would expect.
4. The ambient air temperature is below 32° Fahrenheit, and the smartphone or tablet is too cold. The design operating temperature of iPhones, iPads, and

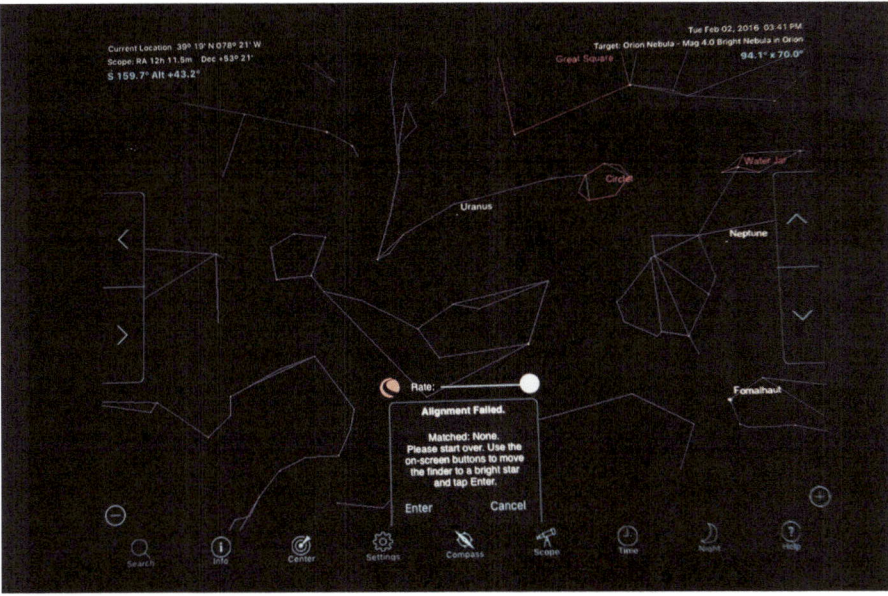

Fig. 4.3 Alignment failed (Chen)

Android phones and tablets is 32–95°F. A high temperatures, the smart device will display a warning message that it is too hot, and to shutdown the device. At extreme low temperatures, the device will cause the Evolution and the SkyPortal to act in strange and mysterious ways. Uncontrollable slewing, "Alignment Failed" messages, and controls that are unresponsive are typical gremlins that indicate the smart device is too cold.

The remedy for the first three of the listed probable causes of alignment failure is summarized in the acronym TOTOTA—Turn Off, Turn On, Try Again. Essentially, a reboot to the startup procedure. More often than not, this will work and the observing session can begin.

In the case of cold weather and ambient temperatures below freezing, it is recommended that the smart device be kept warm in some manner. If using SkyPortal on an iPhone or Android phone, keep the phone in a pocket close to the body in order to keep the device warm. Tablets, being somewhat larger, will benefit from a location under a sweater or jacket. Bring the smart device out only when alignment procedures are being executed, or a GoTo search is initiated. Otherwise, keep the smart device under wraps.

Be aware that in extreme cold weather, the Evolution telescope electronics, battery, and lubrication will be stressed. It is not recommended to be operating telescope electronics under single-digit temperatures for extended periods of time. This is not the fault of Celestron, Apple, or the various Android manufacturers. These are commercial-grade electronics designed for the consumer market and priced as such. Military-grade electronics are designed to operate in temperature ranges as cold as −65 °F and as hot as +125 °F. NASA-grade operating temperatures are rated to even greater extremes. The same applies for military-grade and NASA-grade mechanicals and lubricants. Military-grade and NASA-grade electronics are considerably more expensive than consumer-grade electronics, and are not be affordable to the consumer. The same applies for the mechanicals and lubricants. So, no complaints, please!

When all goes well in the SkyAlign process, the screen in Fig. 4.4 will appear. The user can now progress to conducting GoTo searches.

Optional Automatic Alignment Using StarSense Accessory

The StarSense Telescope Alignment Accessory from Celestron (see Chap. 9, Fig. 9.31) is an integrated digital camera that attaches to the telescope's optical tube in place of the finderscope or red dot finder, and the included StarSense hand controller that connects to compatible computerized Celestron telescope mounts. StarSense is also compatible with the latest version of the SkyPortal app. The StarSense technology uses a built-in camera to automatically identify calibration stars and enable go-to alignment of the telescope's optical tube with celestial objects in the SkyPortal database. The camera automatically captures a series of images of the sky. StarSense identifies the stars in the images, matching them to its

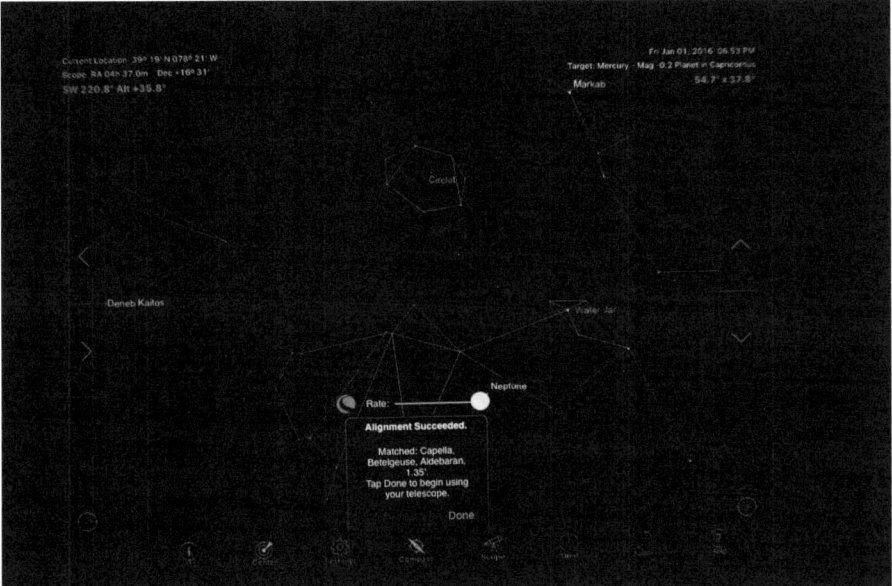

Fig. 4.4 Alignment successful screen for SkyAlign (Chen)

database. Once a positive match is confirmed, StarSense calculates the coordinates of the center of the captured image, thereby determining exactly where the telescope is pointed. After installing the camera and SkyPortal auto-detects its presence, the StarSense accessory collects information from its field of view and delivers precise go-to pointing within 3 min.

Prior to December 23, 2015, the StarSense technology was incompatible with SkyPortal, and could only be used with the StarSense hand controller on NexStar Evolution telescopes. However, with the update SkyPortal Version 1.5.17, this incompatibility was rectified. SkyPortal now supports SkySense AutoAlign. If equipped with the StarSense AutoAlignment accessory, a NexStar Evolution and SkyPortal can be aligned with a single tap of the display screen. The new SkyPortal update supports StarSense EQ, Alt-Az, and Wedge alignment, and StarSense manual align. It is recommended that present users of SkyPortal downloaded prior to December 23, 2015 download the latest version 1.5.17 onto the Apple or Android smart device.

If the telescope is equipped with the optional StarSense AutoAlign system, SkyPortal automatically detects StarSense. For the initial use of the optional accessory, the StarSense camera is mounted upon the NexStar Evolution telescope in place of the StarPointer red dot finder. StarSense is then plugged into any one of the Evolution's four AUX ports. Power up the telescope and WiFi connect with the smart device and SkyPortal. Set the Evolution telescope to its Home position by aligning using the index marks on the mount arm. Tap OK to begin auto alignment.

Once the alignment is complete, tap Align and then Calibrate. The star's new position on the camera will be displayed as a set of coordinates, with 640, 480 as the default center. This will be saved in SkyPortal for future alignments. Once the camera is calibrated, SkyPortal will prompt to start a new StarSense Auto alignment. This procedure occurs for the initial use of StarSense, with all subsequent alignments occurring automatically.

Basic GoTo Search Operations

There are two types of searches that can be conducted in SkyPortal. Before conducting any GoTo search, make sure to insert into the Celestron diagonal attached to the visual back of the telescope a low power eyepiece. The NexStar Evolution comes standard equipped with a 40 mm Plossl eyepiece that is suitable for this task. Many well-heeled Evolution users will have substituted a 2-in. diagonal in order to use the wider field-of-view 2-in. eyepieces. Using wide field-of-view eyepieces will be of aid is GoTo searches.

The first technique is to access the various catalogs stored in SkyPortal for deep sky objects. SkyPortal does not display a deep sky object on its Sky Chart at its normal scale. The object can be located if the screen is zoomed-in sufficiently or found in a GoTo search. Only then is the DSO displayed on the SkyPortal map. Therefore, the second technique, the tap-and-search method, is a challenge for DSO's.

The second technique is simply tapping the fingertip on a star, thus placing crosshairs on it, and then tapping GoTo on the command line of SkyPortal. This works well with GoTo searches for stars.

Searching from the Common Objects List

By tapping the magnifying glass icon marked "Search", in the lower left-hand corner of the Menu Bar located at the bottom of the screen, the Common Objects Lists is brought up and displayed. Fifteen categories of GoTo objects are displayed (Fig. 4.5a, b):

1. Tonight's Best
2. Sun & Planets
3. Moons
4. Asteroids
5. Comets
6. Satellites
7. Named Stars
8. Bright Stars
9. Nearest Stars
10. Double Stars

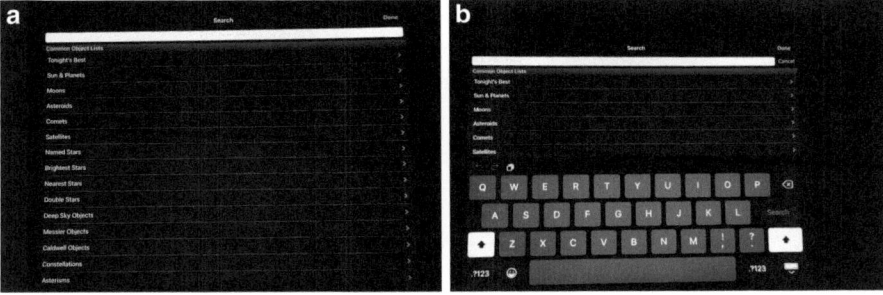

Fig. 4.5 (**a**) Common objects lists (Chen) (**b**) Common objects lists direct search input screen (Chen)

11. Deep Sky Objects
12. Messier Objects
13. Caldwell Objects
14. Constellations
15. Asterisms

By tapping on any one of these categories, the user is presented with a list of objects from which a GoTo search can be accomplished.

If the object name is already known by the user, in the top line the input of the object name or identifier. Tap the search line, and the keyboard will appear to allow input of the object. Type in the object name or catalog identification number (i.e. Orion Nebula or M42) and tap Search.

When selected, a Tonight's Best list is displayed by SkyPortal with an extensive selection of planets, stars, double stars, variable stars, open and globular clusters, nebulae, planetary nebulae, and galaxies. The Tonight's Best List is for those experienced Celestron users who are comfortable with the NexStar+ hand controller Sky Tour. The objects visible for that specific observation time are highlighted, with the remaining objects available for observation later on in the night. As it gets later into the night, the highlighted list will update as celestial objects set and new ones rise (Fig. 4.6).

When an object is selected by tapping the screen over the named object, an Object Info screen will be displayed. The left side of the Object Info screen will contain the following information:

1. Basic Information—Name, catalog number, description, visual magnitude, apparent size, and distance.
2. Visibility—when the object rises, transits, and sets
3. Celestial Coordinations—Azimuth, Altitude, Right Ascension, and Declination
4. Physical Parameters—Color Index, Morphology, Absolute Magnitude, and diameter

The right side of the object info screen gives a text description of the objects, history of discovery, and thorough discussion of the object. This section is often

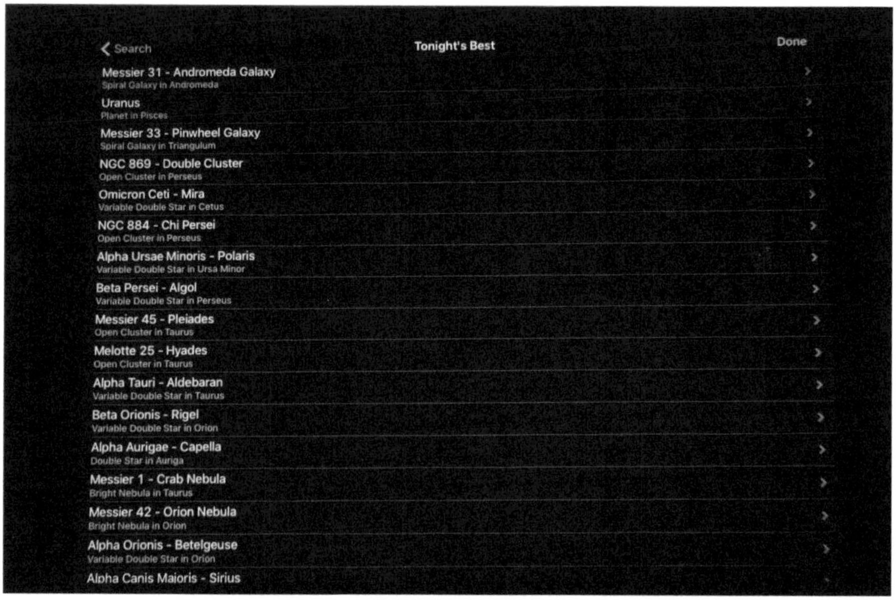

Fig. 4.6 Tonight's best list (Chen)

times accompanied with astrophotos of the object taken from the Hubble Space Telescope or Earth-based telescopes. If the smart device is connected simultaneously (a capability to be discussed in a later chapter) with the Internet, additional photos are sometimes available for display.

As seen in Fig. 4.7a, in the upper left-hand corner of the Object Info screen is an speaker icon marked Celestron Audio. When tapped, this will activate an audio recording describing the object and its historical and scientific significance. The audio recording is similar to that provided by an earlier and no-longer available Celestron Sky Scout handheld planetarium device. Celestron Audio is not available on all objects, but most of the important and significant objects are covered by this unique feature.

All this information is well and good, but to actually perform a GoTo search, the next step is to tap the GoTo icon in the lower left hand corner of the Object Info screen, or the word "Done" in the upper right hand corner of the Object Info screen, returning the screen to the Sky Chart screen and tap GoTo from there (Fig. 4.7b, c).

In the upper right hand corner of the Status Bar is displayed the Target for the GoTo. By tapping GoTo on the command line, the NexStar Evolution will slew to the target object, and when the search is completed, the object will be displayed on the SkyPortal screen with a crosshairs-and-target symbol over the object symbol. A look through the eyepiece will show the object within the field-of-view of the low power eyepiece. When the SkyAlign process is performed properly, every GoTo will result in the same spot in the eyepiece field-of-view (Fig. 4.8).

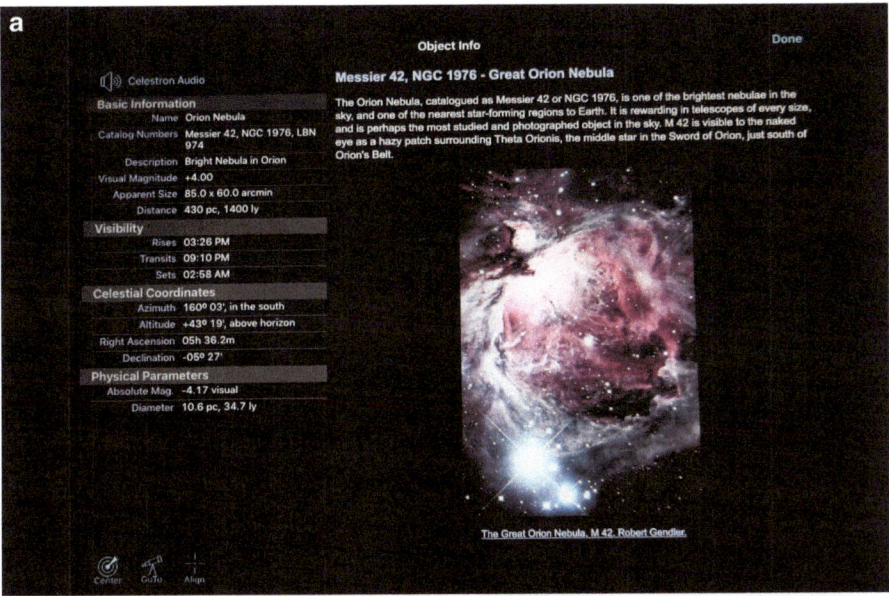

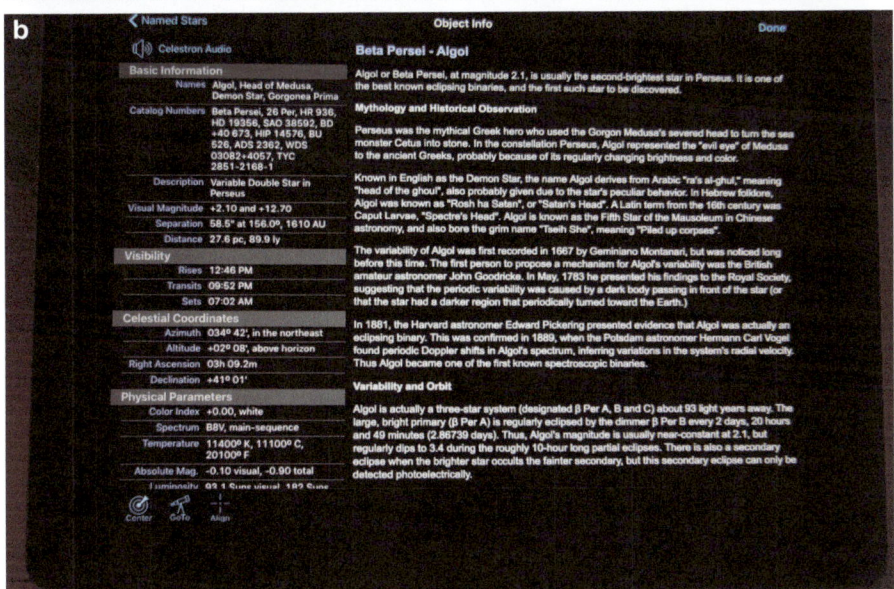

Fig. 4.7 (**a**) Object information screen for M42 The Great Orion Nebula (Chen) (**b**) Object information screen for Algol (Chen)

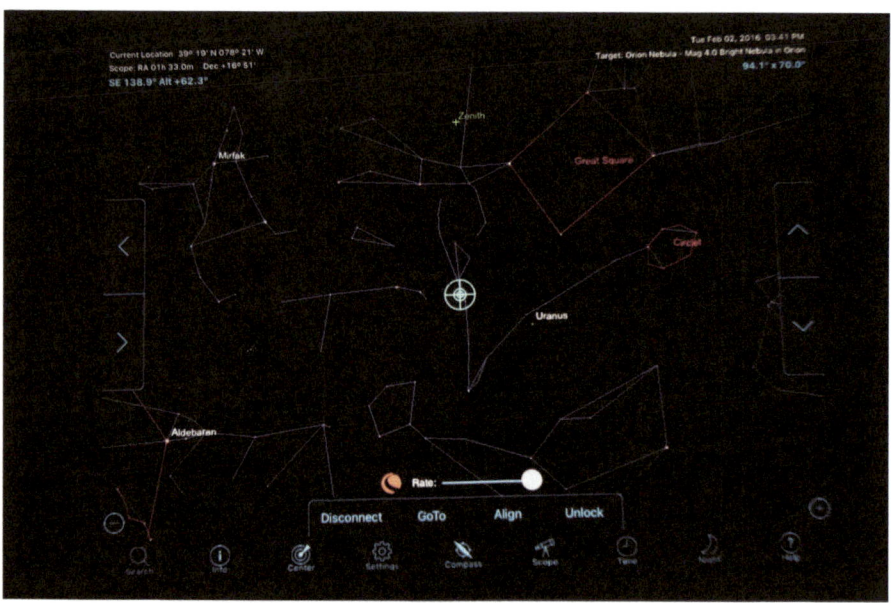

Fig. 4.8 Target sky chart screen (Chen)

The next selection on the Common Objects List is Sun & Planets. As in all the objects lists, the highlighted objects are those visible at the time of observation.

CAUTION: DO NOT OBSERVE THE SUN WITHOUT PROPER SOLAR FILTERING EQUIPMENT. PLEASE READ THE FOLLOWING

1. *Never look directly at the Sun with the naked eye or with a telescope, unless the proper solar filter is being used.* Permanent and irreversible eye damage will result without proper protection.
2. *Never use the telescope to project an image of the Sun onto any surface.* Internal heat build-up can damage the telescope and any accessories attached to it.
3. *Never use an eyepiece solar filter or a Herschel wedge with the Evolution telescope.* Internal heat build-up inside the telescope can cause these devices to crack or break, allowing unfiltered sunlight to pass through to the eye and cause irreparable damage and blindness.
4. *Never leave the telescope unattended when viewing the Sun.* People and children unfamiliar with the dangers of viewing the unfiltered Sun may do something stupid if left alone with the telescope. Never underestimate the dumbness and stupidity of the general public.

Now that this solar warning has been provided, the Sun & Planets list is handled in the same manner as the Tonight's Best list. Tap the desired object, the Object Info screen will be displayed. The next step is to tap the GoTo icon in the lower left hand corner of the Object Info screen, or the word "Done" in the upper right hand corner

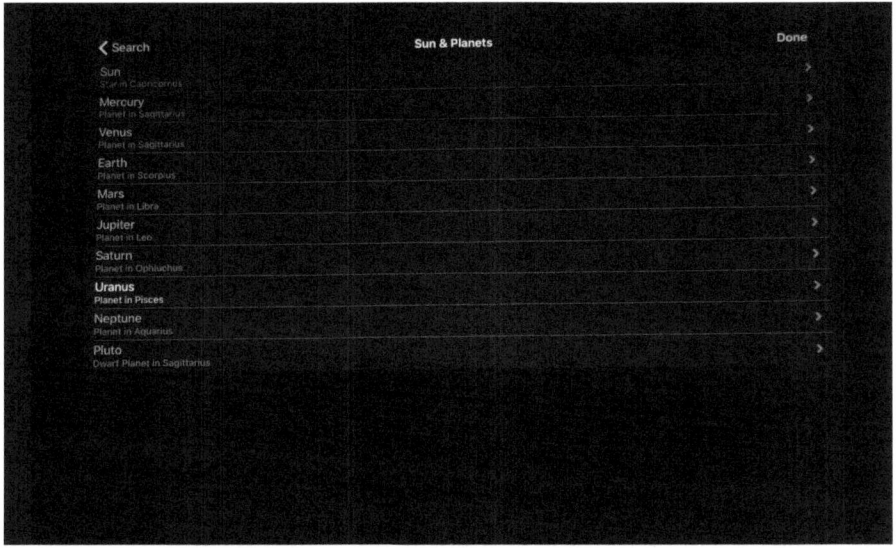

Fig. 4.9 Sun & planets screen (Chen)

of the Object Info screen, returning the screen to the Sky Chart screen. The Sky Chart will then be displayed, with the Target displayed in the upper right hand corner of the Status Bar. Tap GoTo on the command line. The telescope will then slew to the planetary object and observing can now begin.

Observing the Sun is an opportunity to utilize the Hibernate Enabled setting in the Settings menu. This will be discussed in greater detail later in this book. The strategy for Solar observing is to set the telescope during dark nighttime hours, put the telescope into Hibernate Enabled, shutdown the electronics, but don't move the telescope or loosen the clutches. Re-energize the telescope during the daytime for observing the Sun, attach all necessary solar filtering, perform a GoTo search for the Sun using the Sun & Planets menu, and begin observing our home star the Sun. Simple (Fig. 4.9).

The next item of the Common Objects List is Moons (sorry, somewhat awkward but grammatically correct!).

The Moons list is handled in the same manner as the Tonight's Best list. Tap the desired moon, the Object Info screen will be displayed. The next step is to tap the GoTo icon in the lower left hand corner of the Object Info screen, or the word "Done" in the upper right hand corner of the Object Info screen, returning the screen to the Sky Chart screen. The Sky Chart will then be displayed, with the Target moon is displayed in the upper right hand corner of the Status Bar. Tap GoTo on the command line. The telescope will then slew to the desired moon and observing can now begin.

To many readers, this particular list is somewhat odd. Earth's Moon is the only object on the list that presents a target that is highly observable and detailed. The

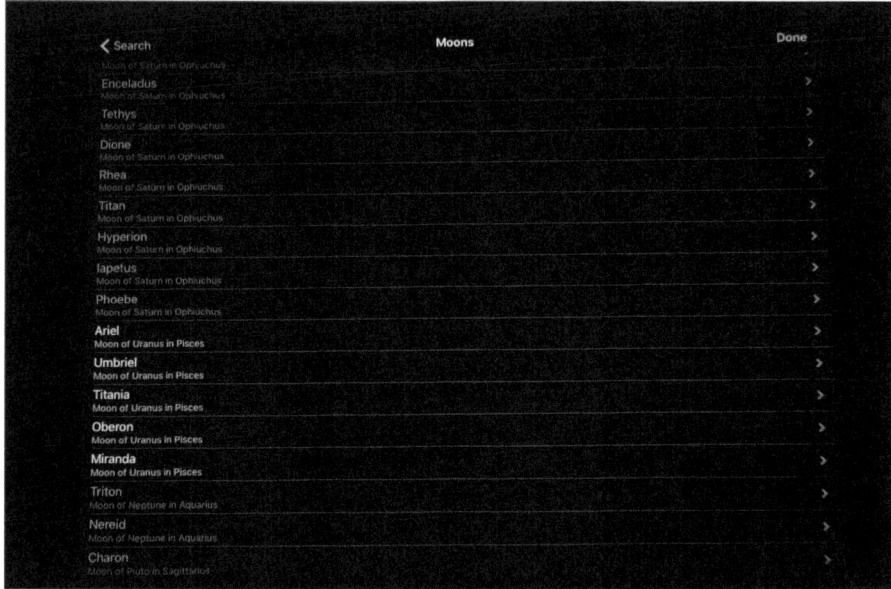

Fig. 4.10 Moons screen (Chen)

remaining moons listed are interesting points of light orbiting their respective planets, and in the case of Jupiter and Saturn's moon, their identification visually is problematic. Many of these moons are observable, but sometimes not easily identifiable (Fig. 4.10).

The Asteroids list is handled in the same manner as with previous lists. Tap the desired asteroid, the Object Info screen will be displayed. The next step is to tap the GoTo icon in the lower left hand corner of the Object Info screen, or the word "Done" in the upper right hand corner of the Object Info screen, returning the screen to the Sky Chart screen. The Sky Chart will then be displayed, with the Target asteroid name displayed in the upper right hand corner of the Status Bar. Tap GoTo on the command line. The telescope will then slew to the asteroid object and observing can now begin.

Identifying the selected named asteroid is problematic. Most asteroids are mere pinpricks of light, and only careful observation of the star positions will yield recognition of the sought-after asteroid. A point of light that changes position against the stationary background of stars will be the sought after asteroid. An asteroid closer in to the Earth will likely "move" against the background in less time, thus being perceivable. A dim or slow moving asteroid may take a number of days to change position against the stationary starry backdrop. Often, only astrophotography will reveal the asteroid's presence. Observing asteroids requires patience, keen eyes, and dark skies. Asteroids are not for beginners (Fig. 4.11).

The Comets list is handled in the same manner as with previous lists. Tap the desired object, the Object Info screen will be displayed. The next step is to tap the

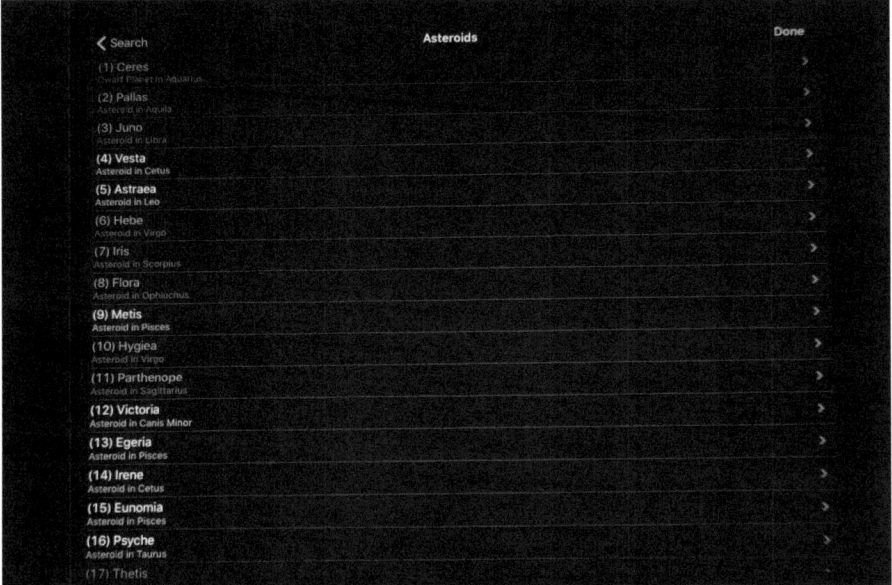

Fig. 4.11 Asteroids screen (Chen)

GoTo icon in the lower left hand corner of the Object Info screen, or the word "Done" in the upper right hand corner of the Object Info screen, returning the screen to the Sky Chart screen. The Sky Chart will then be displayed, with the Target comet name displayed in the upper right hand corner of the Status Bar. Tap GoTo on the command line. The telescope will then slew to the comet and observing can now begin.

Identifying the selected named comet can be problematic. Most comets on the list are dim and not displaying a tail, and are mere pinpricks of light. Only careful observation of the star positions will yield recognition of the sought-after comet. A point of light that changes position against the stationary background of stars will be the sought after comet. A comet closer in to the Earth will likely "move" against the background in less time, thus being perceivable. A dim or slow moving comet may take a number of days to change position against the stationary starry backdrop. Often, only astrophotography will reveal the comet's presence. Observing dim comets requires patience, keen eyes, and dark skies. But when a comet comes close to the Sun and develops a tail or comet head, the observing becomes really fun.

For newly discovered comets, especially those that astronomers predict will be spectacular, look for updates of SkyPortal in the app store. An easy download of the SkyPortal app with updated asteroid and comet lists will be available, with all the orbital elements incorporated into the database (Fig. 4.12).

Choosing the Satellites list is a special case GoTo search selection. A GoTo search will constitute a quick slewing motion to locate the satellite, but expect a fast

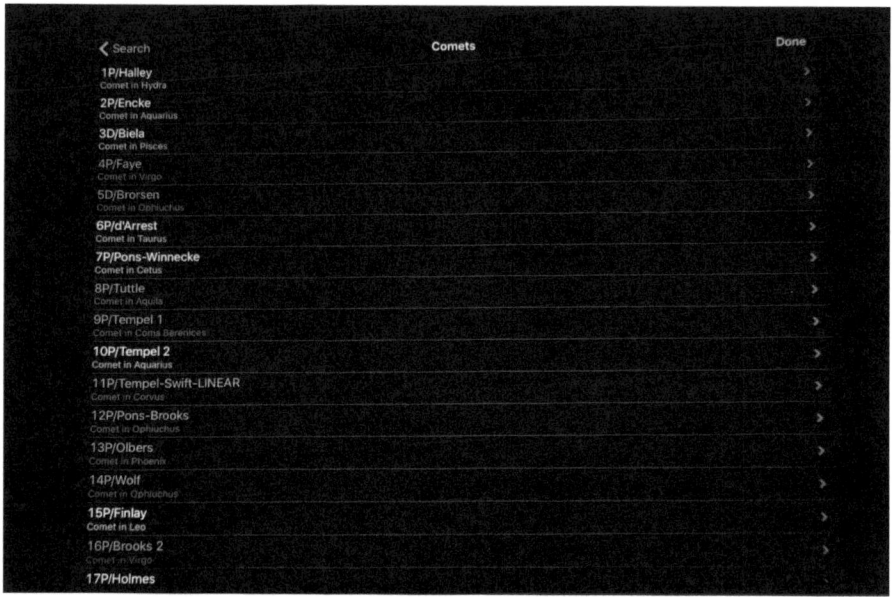

Fig. 4.12 Comets screen (Chen)

slewing motion of the telescope to match the motion of the moving satellite! This means some prior knowledge about a satellite GoTo search is necessary:

1. Clear the area around the telescope to allow free movement for the observer. Remove eyepiece cases, observing chairs, radios, flashlights, etc. that might impede or be a trip hazard.
2. Use a long eye relief eyepiece in the telescope. An eyepiece with 18-20 mm of eye relief will enable the observer to maintain a view of the satellite in the eyepiece as it rapidly moves with the orbital motion of the passing satellite. Objects like the International Space Station traverse the sky at a good pace, and its easy to lose the eyepiece view if the eye relief is too short.
3. Don't be surprised if the satellite is a mere pinpoint of light that is moving. Most satellites are not large enough to image.
4. Don't be surprised if you don't see anything at all. Satellite orbits are dynamic and constantly changing. Orbital elements are continually updated by NASA, ESA, and other space agencies. These updates may not be in SkyPortal database. The most notorious target is the ISS. The ISS is periodically boosted in orbit with maneuvering rocket engines to maintain orbit. The ISS orbital elements change following one of these boosts, and this is not automatically updated in the SkyPortal database. The more recent the SkyPortal update is, the more likely the success of a GoTo and tracking of this sometimes elusive object.

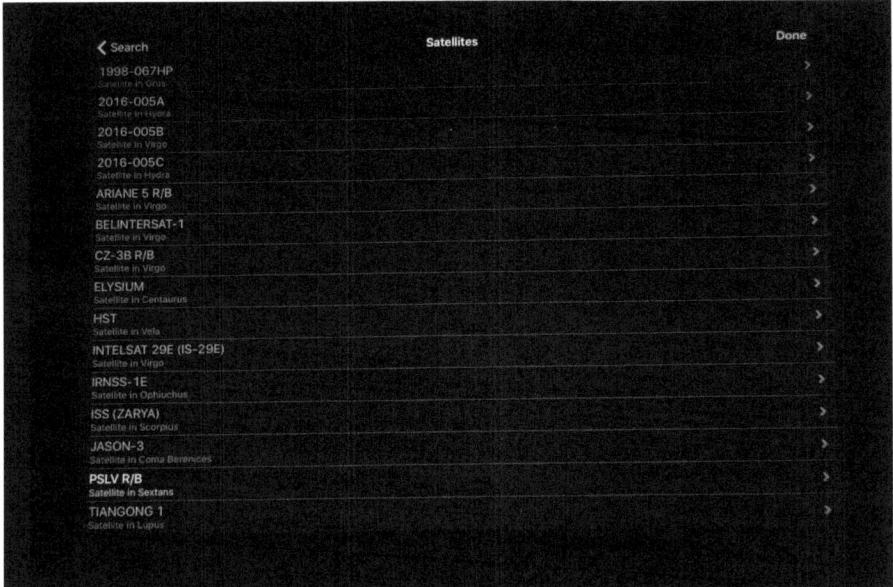

Fig. 4.13 Satellites screen (Chen)

As with the previous lists, selection from the Satellite list is handled in the same manner. Tap the desired satellite, the Object Info screen will be displayed. The next step is to tap the GoTo icon in the lower left hand corner of the Object Info screen, or the word "Done" in the upper right hand corner of the Object Info screen, returning the screen to the Sky Chart screen. The Sky Chart will then be displayed, with the Target satellite name displayed in the upper right hand corner of the Status Bar. Tap GoTo on the command line. The telescope will then slew to the satellite and observing can now begin. Again, don't be surprised that after the GoTo slew is completed that the telescope will continue to move. It is now tracking the motion of the satellite (Fig. 4.13).

The Named Stars list is handled in the same manner as with previous lists. Tap the desired object, the Object Info screen will be displayed. The next step is to tap the GoTo icon in the lower left hand corner of the Object Info screen, or the word "Done" in the upper right hand corner of the Object Info screen, returning the screen to the Sky Chart screen. The Sky Chart will then be displayed, with the Target named star displayed in the upper right hand corner of the Status Bar. Tap GoTo on the command line. The telescope will then slew to the named star and observing can now begin.

This list is quite extensive, and scrolling through the list can be time-consuming. To the right of the Named Stars list is the alphabet listed vertically. The shortcut method of getting through the Named Stars list is tapping the letter of the alphabet that the sought-after star's name begins. For example, tap "P" to jump to the "P" section of the list to locate Polaris (Fig. 4.14).

Fig. 4.14 Named stars screen (Chen)

The Brightest Stars list is handled in the same manner as all of the Common Objects list. Tap the desired object, the Object Info screen will be displayed. The next step is to tap the GoTo icon in the lower left hand corner of the Object Info screen, or the word "Done" in the upper right hand corner of the Object Info screen, returning the screen to the Sky Chart screen. The Sky Chart will then be displayed, with the Target Brightest star displayed in the upper right hand corner of the Status Bar. Tap GoTo on the command line. The telescope will then slew to the selected Brightest star and observing can now begin.

The Brightest Star list contains a listing of first Magnitude stars, that even in bright sky polluted suburbia can be seen through the gray haze of street lights, parking lot lights, and neighborhood porch lights (Fig. 4.15).

The Nearest Stars list is handled in the same manner as with previous lists. Tap the desired object, the Object Info screen will be displayed. The next step is to tap the GoTo icon in the lower left hand corner of the Object Info screen, or the word "Done" in the upper right hand corner of the Object Info screen, returning the screen to the Sky Chart screen. The Sky Chart will then be displayed, with the Target nearest star displayed in the upper right hand corner of the Status Bar. Tap GoTo on the command line. The telescope will then slew to the desired nearest star and observing can now begin.

The Nearest Star list comprises of all the stars in the neighborhood of Earth. Many will be bright and can be seen with the unaided eye, such as Sirius or Alpha

Fig. 4.15 Brightest stars screen (Chen)

Centauri. The bright examples will also be found in the Brightest Stars listing. Some are very dim and can be a challenge to spot even with the light-gathering ability of the NexStar Evolution telescope, such as UV Ceti and BL Ceti. Many of these Nearest Stars are double stars and can also be found in the Common Objects List selections for Double Stars (Fig. 4.16).

The Double Stars list is handled in the same manner as with previous lists. Tap the desired object, the Object Info screen will be displayed. The next step is to tap the GoTo icon in the lower left hand corner of the Object Info screen, or the word "Done" in the upper right hand corner of the Object Info screen, returning the screen to the Sky Chart screen. The Sky Chart will then be displayed, with the Target Double star displayed in the upper right hand corner of the Status Bar. Tap GoTo on the command line. The telescope will then slew to the selected double star and observing can now begin.

Double star observing is a great passion among many amateur astronomers. Many find the splitting a pair of closely separated double stars challenging and rewarding. Splitting closely separated doubles has long served as a test of telescope optics, challenging the capability and quality of both telescope optics and eyepiece design. The SkyPortal list is very comprehensive, offering both the easy and colorful famous doubles plus the close and challenging doubles.

Pick clear and steady nights for splitting doubles. Clear and clean optics are beneficial. When encountering difficulty in splitting a close pair, a good trick is to remove the telescope diagonal, insert the eyepiece into the optic back of the

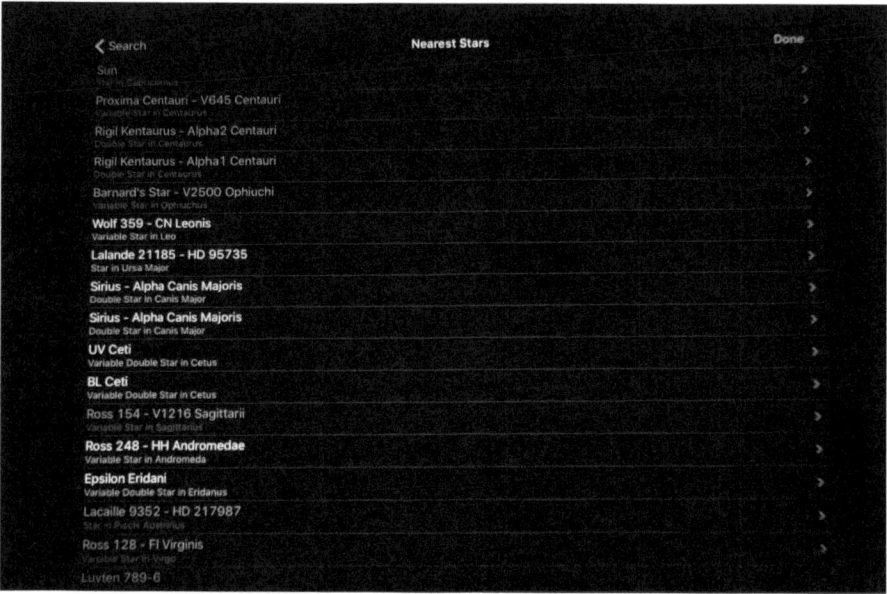

Fig. 4.16 Nearest stars screen (Chen)

NexStar Evolution, and observe the double star using a straight through approach. By removing one layer from the optical chain can be a little uncomfortable, but also can be rewarding (Fig. 4.17).

At the risk of being repetitive, the Deep Sky Objects list is handled in the same manner as with previous lists. Tap the desired object, the Object Info screen will be displayed. The next step is to tap the GoTo icon in the lower left hand corner of the Object Info screen, or the word "Done" in the upper right hand corner of the Object Info screen, returning the screen to the Sky Chart screen. The Sky Chart will then be displayed, with the Target Deep Sky Object displayed in the upper right hand corner of the Status Bar. Tap GoTo on the command line. The telescope will then slew to the selected Deep Sky Object and observing can begin.

All of these objects can be accessed via the Messier or Caldwell Objects lists. Be aware that this Deep Sky Objects lists uses the alternative names for many of objects. For instance, NGC 3242 is more commonly known as the Ghost of Jupiter, but in this listing is referred to by its alternative nomenclature as the CBS Eye. SkyPortal will display alternative names for many objects in the Messier, Caldwell, and Deep Sky Objects listings. Don't be confused. This is just a good way of learning the alternative names and the Messier, Caldwell, or NGC nomenclatures. In fact, M27 is listed twice in the Deep Sky Objects list, once as the Apple Core Nebula and also as the more familiar Dumbbell Nebula. The aforementioned NGC 3242 is also listed under its alternative name of Ghost of Jupiter (Fig. 4.18).

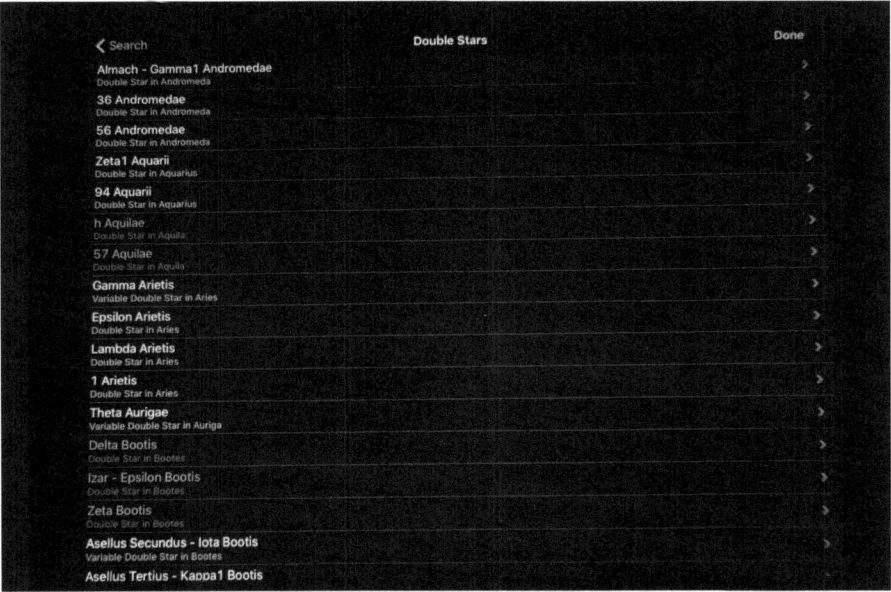

Fig. 4.17 Double stars screen (Chen)

Fig. 4.18 Deep sky objects screen (Chen)

The Messier list is handled in the same manner as with previous lists. Tap the desired object, the Object Info screen will be displayed. The next step is to tap the GoTo icon in the lower left hand corner of the Object Info screen, or the word "Done" in the upper right hand corner of the Object Info screen, returning the screen to the Sky Chart screen. The Sky Chart will then be displayed, with the Target Messier object displayed in the upper right hand corner of the Status Bar. Tap GoTo on the command line. The telescope will then slew to the selected Messier Object and observing can begin.

All the Messier objects are listed here in their numerical order from 1 to 110, with their common name listed when applicable. Unlike the previous Deep Sky Objects, the commonly used name is used in the Messier Objects list. All objects appear only once in this listing.

Of note is the SkyPortal handling of Messier 102. Historically, M102 is one of the "missing" Messier objects that some controversy exists over its true identity. Many astronomers believe that M102 is just a duplication of the Messier observation of M101. The developers of SkyPortal have chosen to use Owen Gingerich's identification of NGC 5866 as the mysterious M102, noting that NGC 5866 closely matches Pierre Mechain's object description in the 1781 printed version of the Messier Catalog, and the object position listed by Charles Messier in his handwritten notes (Fig. 4.19).

The Caldwell Objects list is handled in the same manner as with previous lists. Tap the desired object, the Object Info screen will be displayed. The next step is to

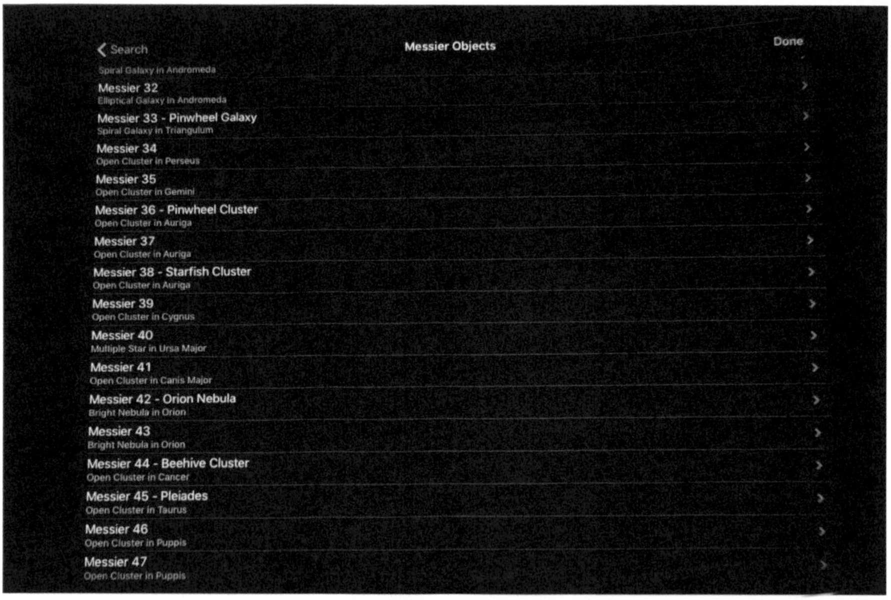

Fig. 4.19 Messier objects screen (Chen)

tap the GoTo icon in the lower left hand corner of the Object Info screen, or the word "Done" in the upper right hand corner of the Object Info screen, returning the screen to the Sky Chart screen. The Sky Chart will then be displayed, with the Target Caldwell Object displayed in the upper right hand corner of the Status Bar. Tap GoTo on the command line. The telescope will then slew to the selected Caldwell Object and observing can begin.

All the Caldwell objects are listed here in their numerical order from 1 to 109, with their common name listed when applicable. Unlike the previous Deep Sky Objects, the commonly used name is used in the Caldwell Objects list. All objects appear only once in this listing.

While the Messier Catalog is used by amateur astronomers as a list of deep-sky objects for observation, Sir Patrick Moore noted that Messier's list did not include many of the sky's brightest deep-sky objects, such as the open cluster Hyades, the h and chi Perseus pair of clusters known as the Double Cluster NGC 869 and NGC 884, and the spiral galaxy in Sculptor NGC 253. Moore also observed that since Messier compiled his list from observations in Paris, it did not include bright deep-sky objects visible in the Southern Hemisphere, such as NGC 5139 Omega Centauri, NGC 5138 Centaurus A, NGC 4755 the Jewell Box and NGC 104 47 Tucanae. Those living in the northern regions will not be able to observe the bright Southern Hemisphere Caldwell Objects from their backyard. A trip to observing locations below latitude 30° will provide the opportunity to observe these Caldwell objects (Fig. 4.20).

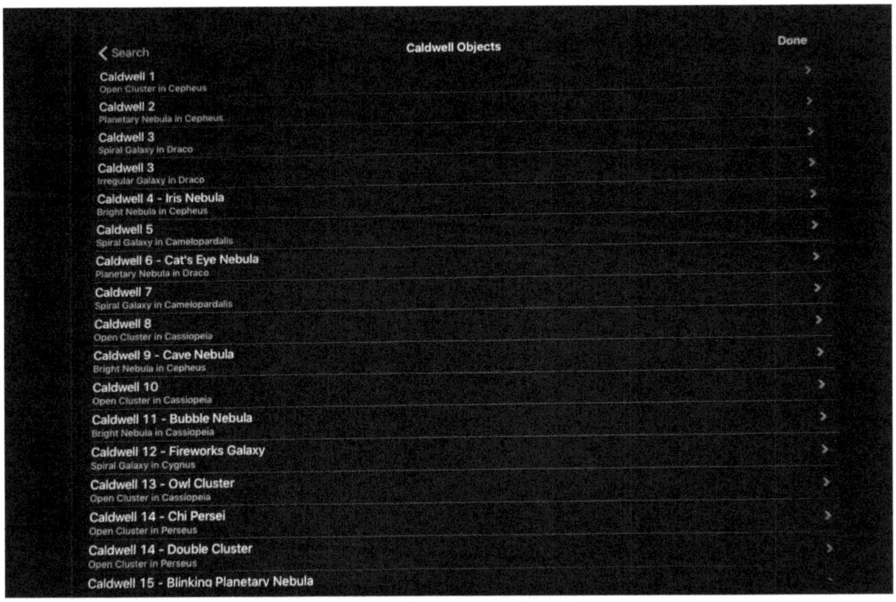

Fig. 4.20 Caldwell objects screen (Chen)

Fig. 4.21 Constellations screen (Chen)

The Constellations list is handled in the same manner as all of the Common Objects lists. Tap the desired object, the Object Info screen will be displayed. The next step is to tap the GoTo icon in the lower left hand corner of the Object Info screen, or the word "Done" in the upper right hand corner of the Object Info screen, returning the screen to the Sky Chart screen. The Sky Chart will then be displayed, with the Target Constellation displayed in the upper right hand corner of the Status Bar. Tap GoTo on the command line. The telescope will then slew to the selected Constellation and observing can begin.

The Constellations menu is, at first glance, somewhat odd. Telescopes are not needed to sight constellations due to their large angular size. The Constellation function is useful when learning the constellation patterns and their locations in the sky. The Evolution/SkyPortal combo will point the telescope in the direction of the sky where the constellation is located, thus providing direction to the naked-eye observer (Fig. 4.21).

The Asterisms list is handled in the same manner as all of the Common Objects lists. Tap the desired object, the Object Info screen will be displayed. The next step is to tap the GoTo icon in the lower left hand corner of the Object Info screen, or the word "Done" in the upper right hand corner of the Object Info screen, returning the screen to the Sky Chart screen. The Sky Chart will then be displayed, with the Target Asterism displayed in the upper right hand corner of the Status Bar. Tap GoTo on the command line. The telescope will then slew to the selected Asterism and observing can begin.

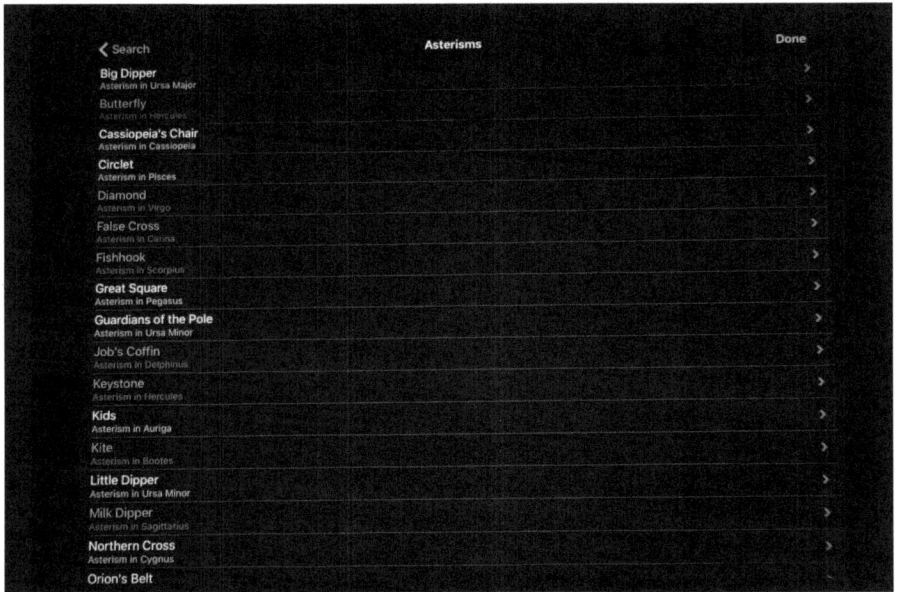

Fig. 4.22 Asterisms screen (Chen)

The Asterisms list is a listing of common named objects, such as the Big Dipper (Ursa Major), Great Square (Pegasus), Northern Cross (Cygnus), and the Little Dipper (Ursa Minor).

An interesting note, as complete as the alternative names are for Messier and Caldwell objects are provided in the description information, alternative names for asterisms are not provided. For instance, the Big Dipper is also known as the Plow is some European countries (Fig. 4.22).

The Meteor Showers list is handled in the same manner as with previous lists. Tap the desired object, the Object Info screen will be displayed. The next step is to tap the GoTo icon in the lower left hand corner of the Object Info screen, or the word "Done" in the upper right hand corner of the Object Info screen, returning the screen to the Sky Chart screen. The Sky Chart will then be displayed, with the Target Meteor Shower displayed in the upper right hand corner of the Status Bar. Tap GoTo on the command line. The telescope will then slew to the selected Meteor Shower direction and observing can begin.

As with the Constellation list, the Meteor Showers menu is, at first glance, somewhat odd. A telescope is not needed to see a meteor shower due to their large angular size. The Meteor Showers function is useful when learning the general direction in the sky that a meteor shower appears to originate. The Evolution/ SkyPortal combo will point the telescope in the direction of the sky where the meteor shower apparently originates, thus providing direction to the naked-eye observer.

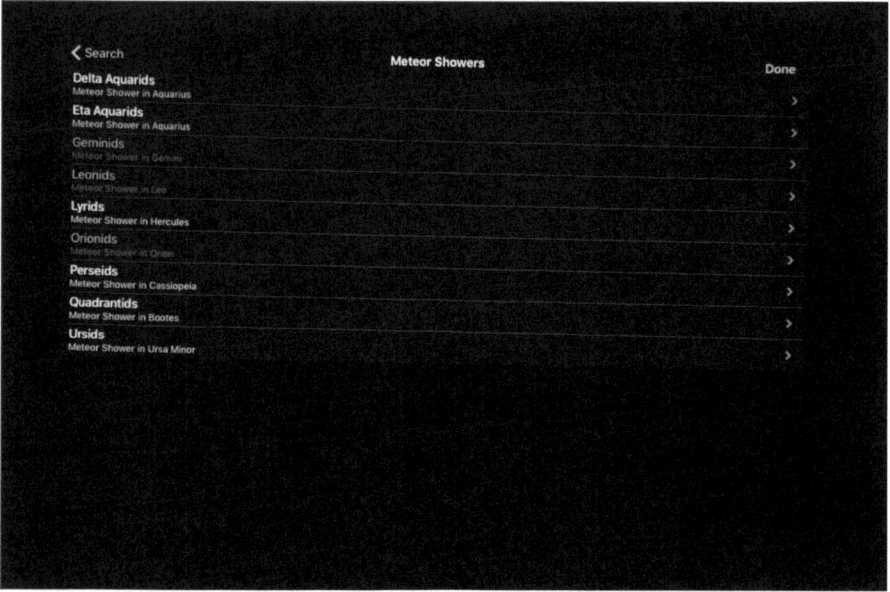

Fig. 4.23 Meteor showers screen (Chen)

By tapping the Info on the Menu bar, information can be accessed on any of the objects being viewed and tracked. The contents of the information screen was discussed earlier in this chapter. In many cases, Celestron Audio is available, which by tapping the speaker icon in the upper left hand location of the information screen will play an audio description of the selected object.

For astronomy outreach, the Celestron Audio can be very handy. By connecting the smart device to some audio speaker accessories that are commercially available, the Celestron audio can be very informative to a wide audience of first-timers.

The lower left hand corner of Fig. 4.23 shows three icons, Center, GoTo, and Align. From the Info screen, using these shortcut commands, the user can place the GoTo object on the center of the Star Chart, initiate a GoTo search, or use the sought object for alignment (Fig. 4.24).

Next on the Menu bar is the icon for Center. Center merely adjusts the Sky Chart screen to display the selected GoTo object in the center of the Sky Chart. This is helpful in keeping oriented with the positioning of the viewed object with the rest of the sky (Fig. 4.25).

By tapping Settings on the Menu bar, the following settings can be managed in SkyPortal:

1. Date & Time
2. Location
3. Appearance

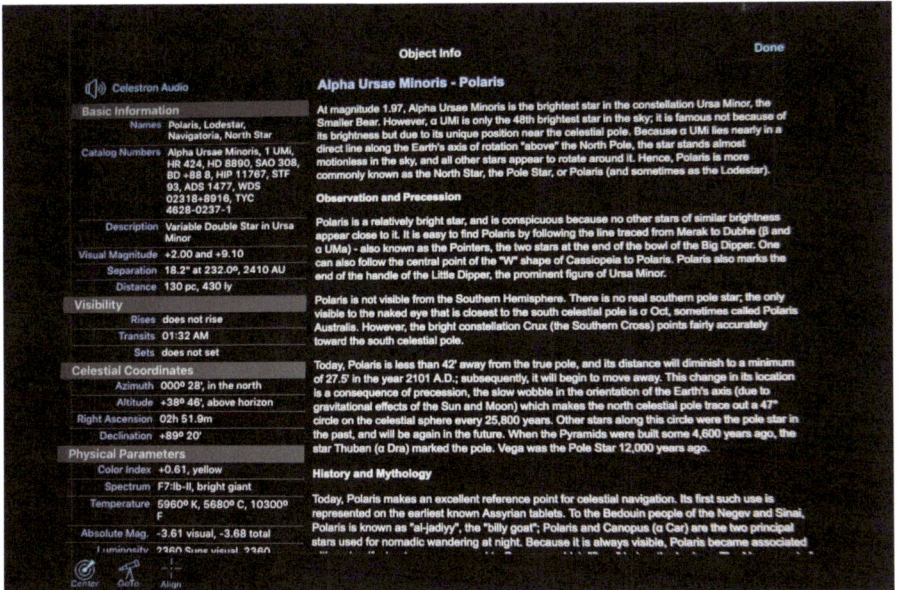

Fig. 4.24 Info screen (Chen)

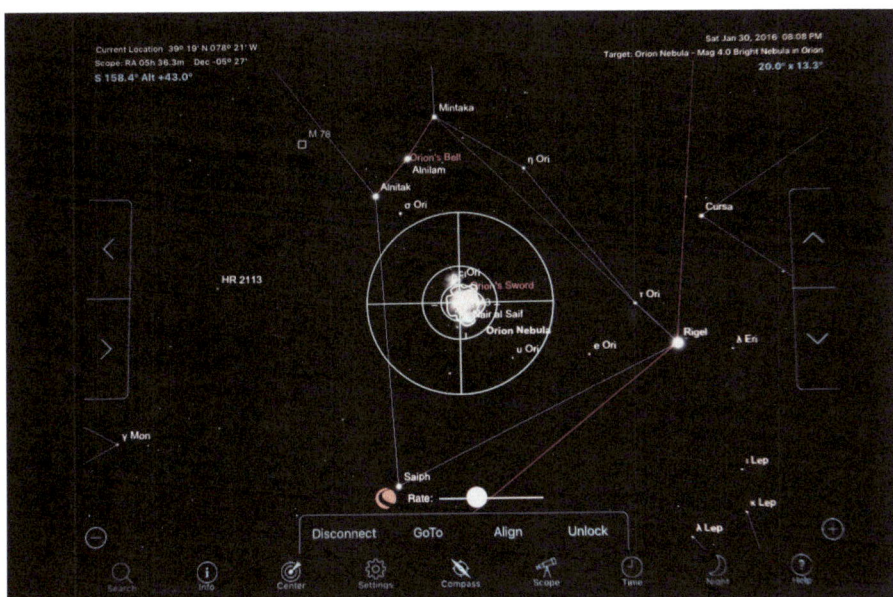

Fig. 4.25 Sky Chart screen with Menu bar icons displayed at the bottom (Chen)

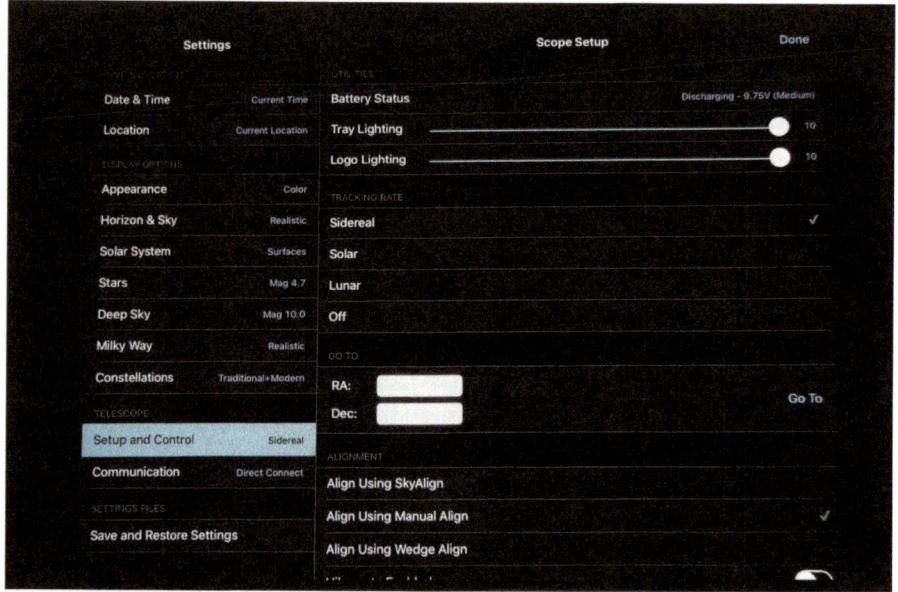

Fig. 4.26 Settings screen (Chen)

4. Horizon & Sky
5. Solar System
6. Stars
7. Deep Sky
8. Milky Way
9. Constellations
10. Setup and Controller Communication
11. Save and Restore Settings (Fig. 4.26)

By tapping the Date & Time, the user can adjust the current date and time for the SkyPortal to use. In reality, unless the user is doing something unique, setting these parameters is not necessary. SkyPortal automatically obtains the date, time, and location from the smart device (Fig. 4.27).

By tapping the Location, the user can adjust the current location for the SkyPortal to use. In reality, setting these parameters is not necessary. SkyPortal automatically obtains the date, time, and location from the smart device. The user is given two other options for location: (1) Choose Location from List and (2) Choose Location from Map. From an accuracy point of view, using the time and coordinates from GPS is the preferred method (Fig. 4.28).

By tapping Appearance, the user can adjust the screen appearance for the SkyPortal screens. The default settings are quite attractive, but depending on the user preference, the chart color and brightness can be changed. The behavior of

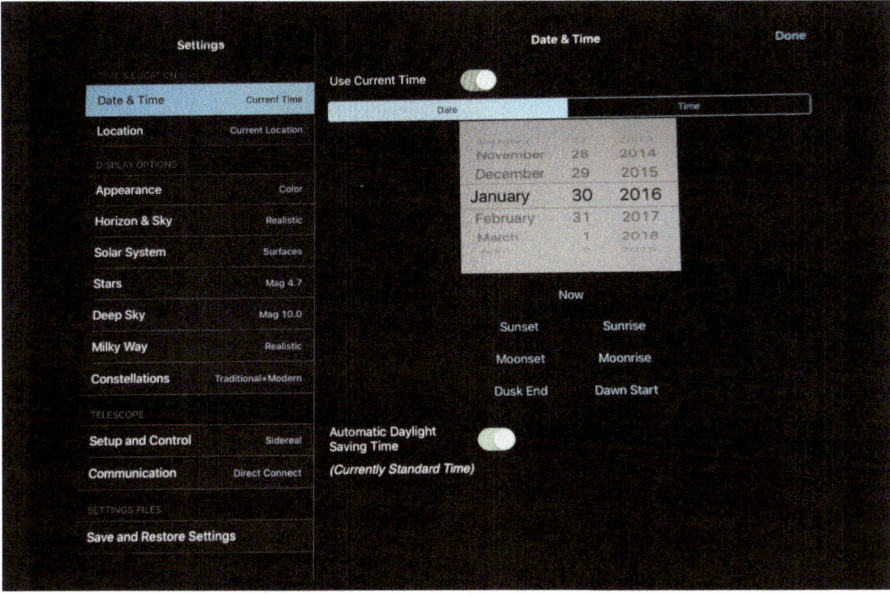

Fig. 4.27 Settings: Date & time screen (Chen)

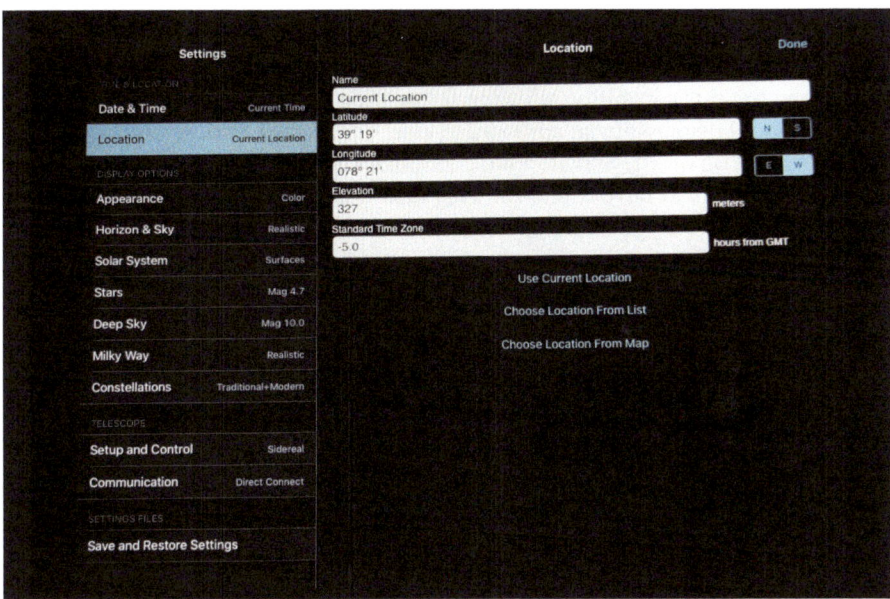

Fig. 4.28 Settings: Location screen (Chen)

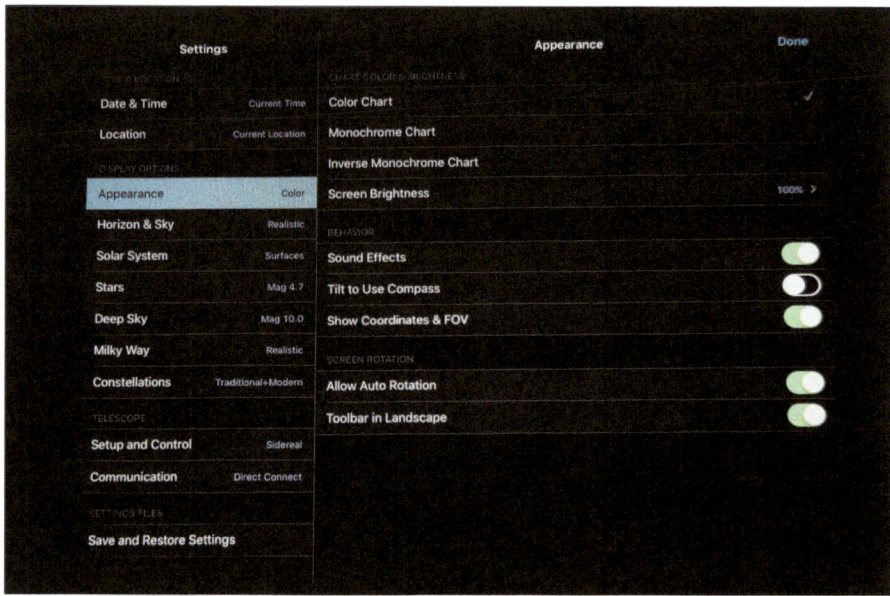

Fig. 4.29 Settings: Appearance screen (Chen)

SkyPortal using the inherent characteristics of the smart device can be adjusted. The use of sound effects can be toggled on or off. Many smart device users are accustomed to sky planetarium apps that take advantage of the built-in compass. SkyPortal has this capability, allowing for holding the smart device against the sky and SkyPortal displaying the appropriate view on the screen for reference. This can as be toggled on or off. Showing the coordinates and field-of-view, allowing auto-rotation of the smart device, and showing the toolbar in the landscape are also user adjustable on or off.

There is a difference between the iOS version of SkyPortal versus the Android version within the Appearance settings. The iOS version of SkyPortal has a Screen Brightness setting, found under Settings > Appearance, which allows the adjustment of the brightness of the screen. This screen brightness applies only to SkyPortal. Other apps running under iOS, and all Android apps including SkyPortal, the screen brightness must be adjusted from the smart device's main Settings menu (Fig. 4.29).

By tapping the Horizon & Sky, the user can adjust the appearance of the background appearance of the Sky Chart for the SkyPortal to use. Here again, the default is acceptable, but the user is afforded the options of how the horizon and sky is depicted on the Sky Chart. SkyPortal provides the background selection of Horizon Panoramas among Mauna Kea, Uluru, or Yerba Buena (Fig. 4.30).

By tapping Solar System, the user can adjust the appearance of the solar system objects for the SkyPortal to use. Options under the Planet & Moon Display include

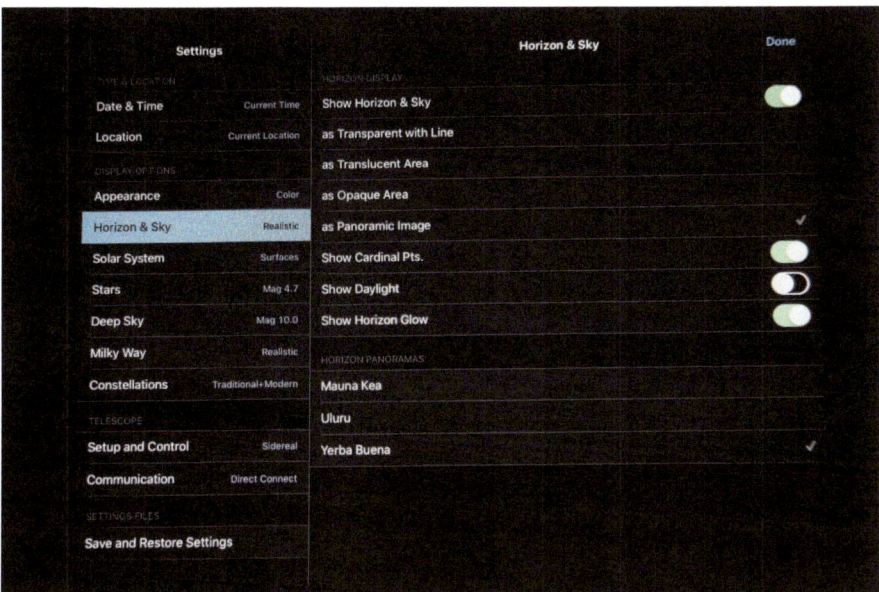

Fig. 4.30 Settings: Horizon & sky screen (Chen)

toggling on or off Show Planets, with Grids, with Planets, with Surfaces, and with Names. Under normal use, SkyPortal does not display planets (other than a small dot) on the Sky Chart until the user "pinches and zooms" the area the planet is located. Zooming in or out can also be accomplished by touching and holding on the + or − sign in the lower right and left corners of the Sky Chart. Under the Minor Body Display options, the user can toggle on or off Show Asteroids, Show Comets, Show Satellites, and with Names. The reader is invited to try each or all of these display options, with the caveat that the Star Chart will become quite cluttered when all display options are toggled to the On position (Fig. 4.31).

By tapping Stars, the user can adjust the appearance of the stars on the SkyPortal Sky Chart. SkyPortal allows the toggling on or off of stars displayed, the magnitude of the stars displayed. The user under the Star Label Display can control the display of the star names, proper names, show the Greek symbols for the stars, and control the name density on the display to avoid clutter. The Star Symbol Options allow for the control of the star symbol size and the symbol intensity.

In practical use, the solar system and star controls are most useful depending on the size of the smart device screen. An iPhone or Android phone screen can quickly fill up with clutter from an excess of information displayed. An iPad or Android tablet can present more information on the display owing to the larger size display, with the large full sized tablets having an advantage over the smaller palm size tablets (Fig. 4.32).

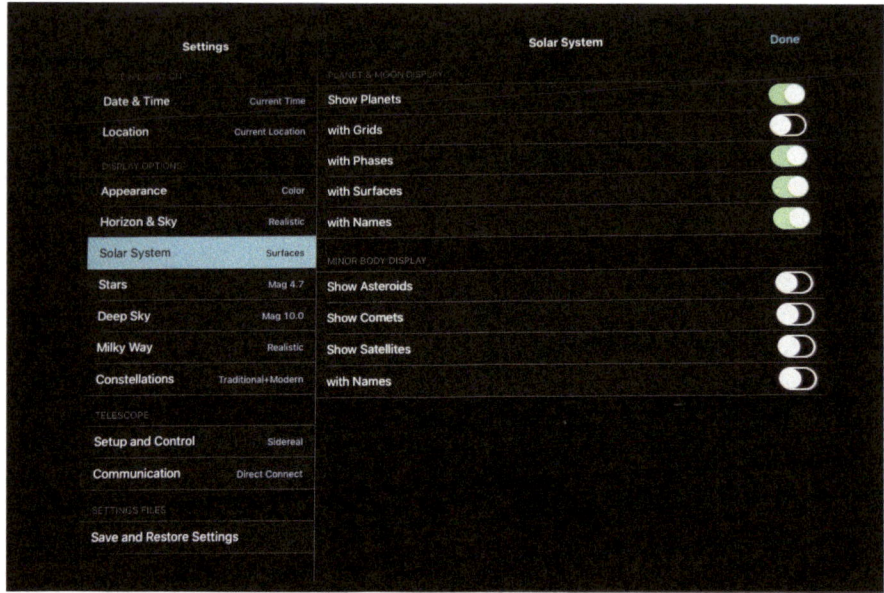

Fig. 4.31 Settings: Solar system screen (Chen)

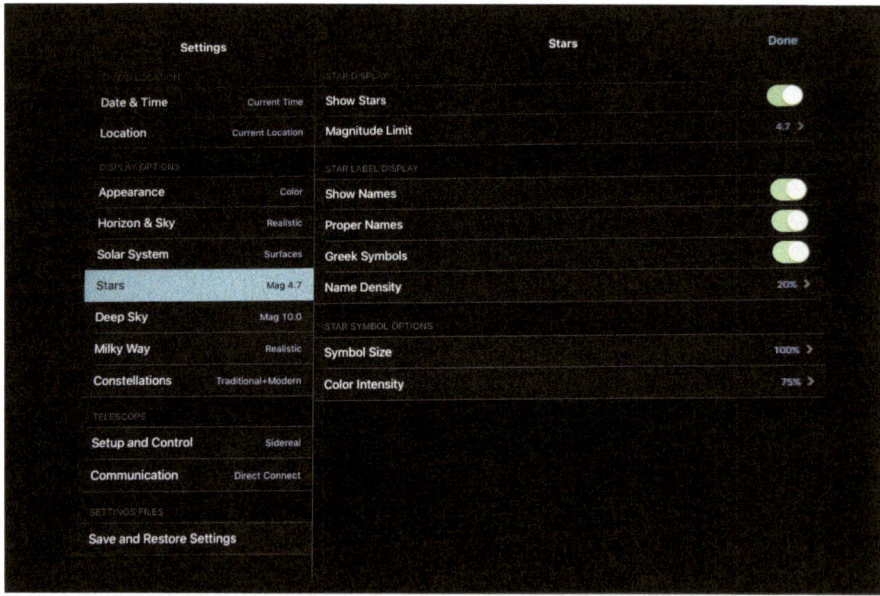

Fig. 4.32 Settings: Stars screen (Chen)

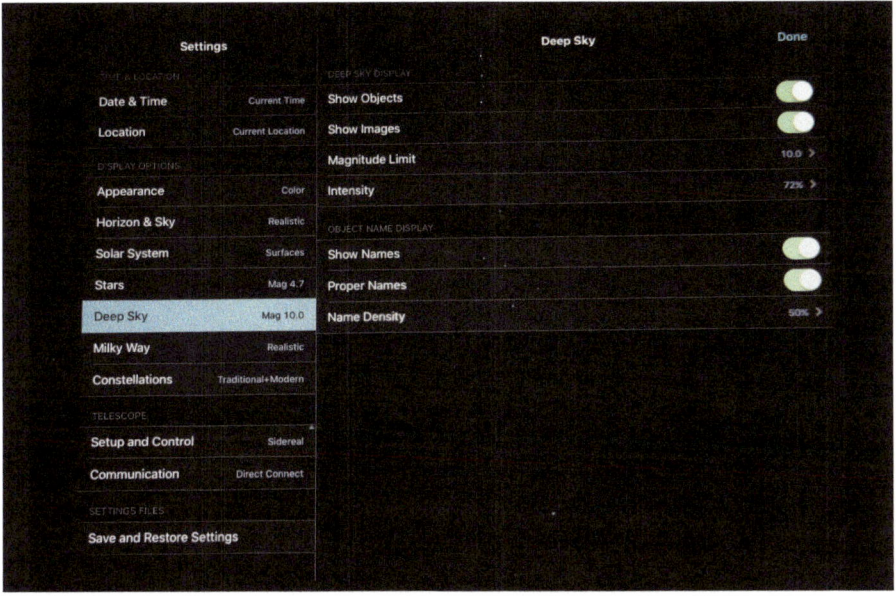

Fig. 4.33 Settings: Deep sky screen (Chen)

By tapping Deep Sky, the user can adjust the appearance of deep sky objects for the SkyPortal to use. Under normal use, SkyPortal does not display deep sky objects on the Sky Chart until the user "pinches and zooms" the area the deep sky object is located. Zooming in or out can also be accomplished by touching and holding on the + or − sign in the lower right and left corners of the Sky Chart. Deep Sky Objects are depicted when a GoTo search has been conducted for the objects, and located on the Sky Chart display with a crosshair target over it. The user can control the Deep Sky display by toggling on or off Show Objects, Show images, set the magnitude limit, and intensity. Object Name Display allows for Show Names, Proper Names, and Name Density (Fig. 4.33).

By tapping Milky Way, the user can adjust the display appearance of the Milky Way for the SkyPortal to use. In reality, this is an attractive display option useful for the user orientation in the night sky. The Milky Way Display controls can toggle on or off the display of the Milky Way, and allows for the display of the Milky Way as a framed outline, filled area, or as a realistic image. Also adjustable is the intensity of the Milky Way depiction and allowing it to fade in smaller field as the user zooms in on the display (Fig. 4.34).

By tapping Constellations, the user can adjust the appearance of the constellations on the SkyPortal Sky Chart display. SkyPortal allows for the display of Constellations as Traditional lines, Modern lines, Mythical Figures, or IAU Boundaries. The user can elect to display the Zodiac Only, adjust the constellation intensity, and toggle Tap to Select. The user can elect to show the constellation names as full names or abbreviations, as well as display asterisms and their names.

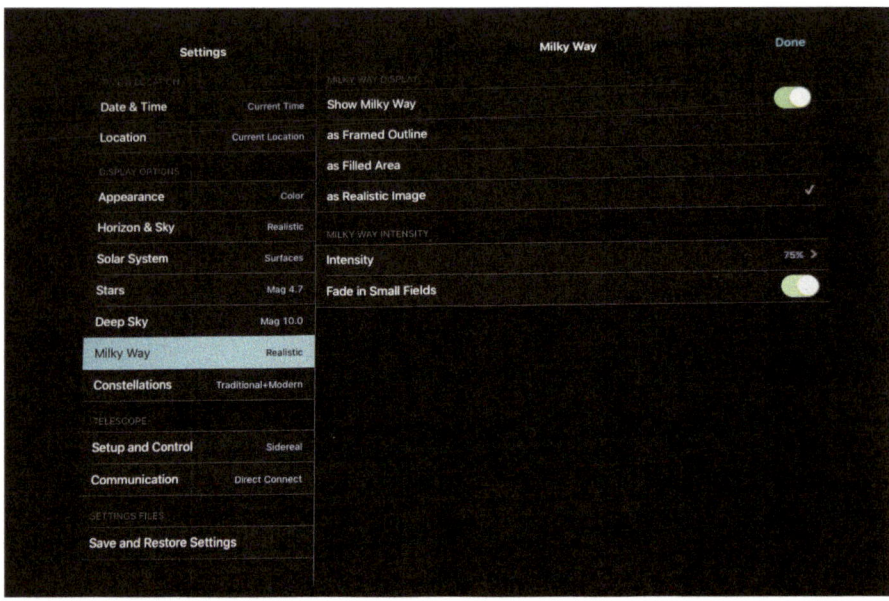

Fig. 4.34 Settings: Milky way screen (Chen)

As with previous controls, in practical use, the constellation controls are most useful depending on the size of the smart device screen. An iPhone or Android phone screen can quickly fill up with clutter from too much information displayed. An iPad or Android tablet can present more information on the display owing to the larger size display, with the large full sized tablets having an advantage over the smaller palm size tablets (Fig. 4.35).

By tapping Setup and Control, the user is afforded a number of settings directed at the NexStar Evolution. Slider controls for adjusting the brightness of the tray light and the Celestron logo light indicators for WiFi connection and power use status are at the top of the right side screen, along with battery status. The choice among sidereal, solar, and lunar tracking is selectable, with a drive turned off option. At the bottom of the screen, SkyPortal offers a selection of alignment options, including SkyAlign, manual align, wedge align for those using the optional WiFi module with a Celestron German equatorial mount, and a hibernate option.

The Setup and Control screen offers several items that require further explanation. There is a RA/Dec GoTo option that allows the user to slew the telescope to objects not listed in the SkyPortal database by entering the right ascension and declination coordinates to slew to a specified area of the sky. This is the quickest way to go to a custom object, such as a new comet (if the RA and Dec coordinates are known) or object of interest from a star catalog (such as NGC/IC objects) or publish reference. Tap the RA or Dec line, and the keyboard will appear to allow input of the object.

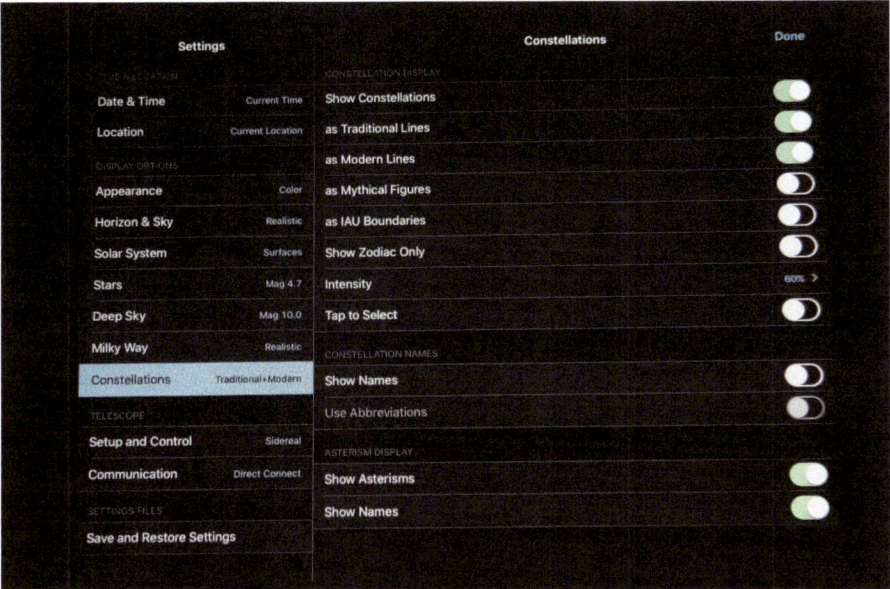

Fig. 4.35 Settings: Constellations screen (Chen)

The Hibernate Enabled allows the telescope alignment to be saved when shutting the telescope off or disconnecting and exiting the SkyPortal app. For Hibernate Enabled to work properly, ALL CLUTCHES must remain engaged and unaltered, and the telescope must not be physically moved. This enables users with permanent or semi-permanent installations to forego re-aligning with every use, or users who are taking a break in observing.

The Altitude Slew Limits are useful for instances where an oversized accessory or a non-standard optical tube assembly is used on the NexStar Evolution mount that may interfere with the mount during the slewing motion. By limiting the altitude slew range, any potential interference and striking of the mount can be avoided. Celestron notes that the slew limit does not work until the telescope is aligned with the sky, or the SkyPortal slew limit assumes the telescope is pointing horizontally.

All mechanical gears have a certain amount of play between the gears known as backlash, even those within the precision made NexStar Evolution. This is seen as a delay in the time it takes for the telescope to move after a direction arrow is engaged. Anti-backlash compensates for backlash by inputting a value which rewinds the motors just enough to eliminate the play between the gears.

The user can toggle on or off Reverse Left and Right, and Reverse Up and Down. The telescope direction can be reversed to change the apparent motion of the star or object in the telescope's eyepiece in the three lowest slew speeds. Reverse Up and Down is the default setting so the star moves in the same direction as the direction button. In practical use, the default setting is preferred.

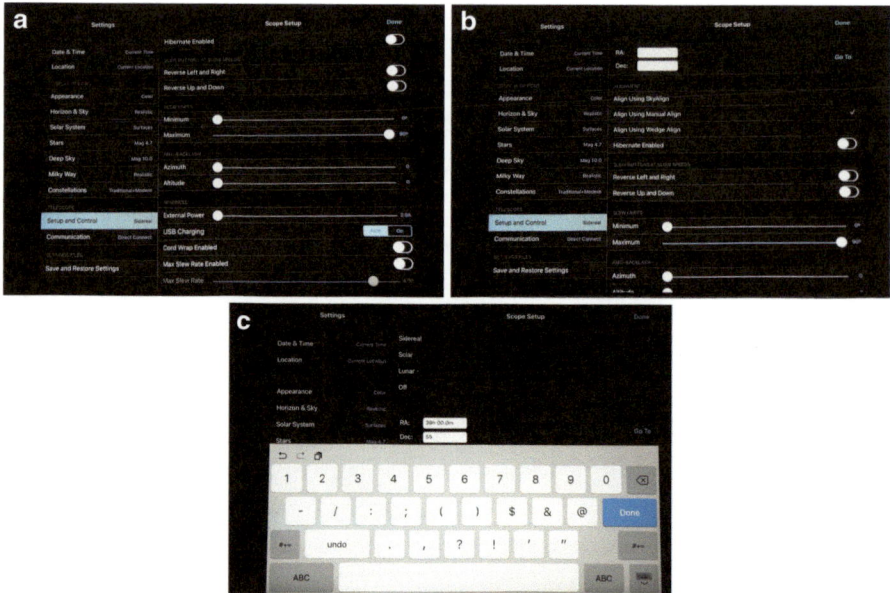

Fig. 4.36 (**a**) Settings: Setup and control screen (Chen) (**b**) Settings: Setup and control screen continued (Chen) (**c**) Settings and control RA and Dec input screen (Chen)

The NexStar Evolution and SkyPortal allows two power management features. The External Power setting adjusts the maximum potential current draw from the power supply. The default setting matches the Celestron supplied 2.0 amp power supply. As Celestron notes: Any setting higher than 2.0 amps requires a higher capacity power supply, which is not supplied with the telescope. The NexStar Evolution has a built-in fail safes if the External Power setting is incorrectly set. The recommendation here is to always use the Celestron supplied power supply. That is what the telescope is designed to use, and the backrooms of many telescope shops are filled with telescopes awaiting repair because of someone's insistence on improperly using a non-standard power supply.

The USB Charger option sets the USB plug on the Evolution mount to always On or Auto. The default setting is Auto, meaning the charger will shut off to save battery life when the battery is low. The On position will force the charger to stay on at all times, even when the battery is low (Fig. 4.36a, b, c).

The Communication options are the Use Direct Connect, and Use Access Point. Direct Connect is the default setting, which allows a direct WiFi interface with the user's smart device. Use Access Point is covered in a later chapter (Fig. 4.37).

If the user has specific settings outside of the default settings, the unique settings can be saved and restored.

To the right of the Settings icon on the Menu bar is the Compass icon. SkyPortal Compass utilizes the internal smart device compass or gyro to center the sky chart

Chapter 5

Basic Operation of the Celestron NexStar Evolution and NexStar+ Hand Control

Included with every Celestron Evolution telescope is the NexStar+ hand control. This is the latest version of the classic Celestron hand control that has been a part of every Celestron NexStar telescope since its introduction.

The question arises, why use the NexStar+ hand control when the SkyPortal app is available? Quite simply, not everyone owns a smartphone or tablet. Additionally, the hand control, and particular the LCD display, will operate in lower temperatures (within reason) where a smartphone or tablet maybe challenged. The NexStar+ HC also represents a backup to the smart device in case of dead battery situations and the like. The NexStar+ HC also has a larger database of deep sky objects, represented by the NGC, IC, Abell, and CCD Objects, than SkyPortal.

For those familiar with the NexStar+ hand control, this chapter will cover familiar territory and can easily be bypassed. Those new to the Celestron NexStar+ hand control will find this chapter useful (Fig. 5.1).

The NexStar+ hand control is the latest iteration of hand held control computers for Celestron GoTo telescopes. In the implementation for the NexStar Evolution series, the NexStar+ hand control is provided as standard equipment, and may be plugged into any of the four AUX on the NexStar Evolution mount.

The architecture of the NexStar+ hand control, from top to bottom, is as follows:

1. Liquid Crystal Display (LCD)—A four line18 character display screen with red backlighting to preserve night vision.
2. Align button—Used during the alignment processing.
3. Direction Keys—Used for directing the direction of slewing motions.

© Springer International Publishing Switzerland 2016

J.L. Chen, *The NexStar Evolution and SkyPortal User's Guide*,
The Patrick Moore Practical Astronomy Series,
DOI 10.1007/978-3-319-32539-2_5

Fig. 5.1 The NexStar+ hand control (Chen)

4. The Numerical+ Catalog Keypad—Dual function keypad. Various on-board catalogs, such as Solar System, Stars, and Deep Sky are accessed using these keys, and as a numerical keypad, objects with numerical nomenclature are identified. Additional functional keys include Identify, Menu, Option, Star Tour, scrolling keys, object info, and Help keys.
5. Option (The Celestron logo key)—Used in combination with other keys to access more advanced features and functions of the NexStar+ hand control.
6. Motor Speed—Used by the user to adjust the slewing speed of the telescope mount.
7. RS-232 jack—not shown in the accompanying photo, this jack is located on the bottom of the hand control, and allows the user to connect the telescope to a computer, for use with software programs for point-and—click slewing capability, and for updating the NexStar+ firmware using a PC.

Alignment

There are a larger number of alignment options available using the NexStar+ hand control (from now on to be referred to as the NexStar+ HC) than available on the SkyPortal app.

1. SkyAlign—the three bright star alignment procedure as used in the SkyPortal app. As with the app, the user is not required to know the name of the star being used for alignment.
2. One Star Align—Utilizes the entered time, date, and location information and allows the user to select a single alignment star. This is a quick procedure that

trades off fast quick setups for accurate GoTo searches. GoTo searches using One Star Align gets the telescope pointing in the neighborhood of the desired object.

3. Two Star Align—Utilizes the entered time, date, and location information and allows the user to select two alignment stars from a list of named alignment stars.

4. Auto Two Star Align—Similar to the aforementioned Two Star Align, but after the first star is selected and aligned, the NexStar+ HC will automatically select a second star for the best possible alignment. Once selected, the telescope will automatically slew to the second alignment star to complete the alignment.

5. Solar System Align—The NexStar+ HC will display a list of visible daytime objects (Planets and the Moon) available to align the telescope. **Please note, any safe observation of the Sun requires the use of proper solar equipment, such as over-the-objective sun filters or solar Hydrogen-alpha filtering. Observing without proper precautions will result in damage to the eyes.**

6. Quick Align—This description of this alignment procedure is still listed in the owner's manual. Older firmware versions prior to Version 4.0 include this procedure. Version 4.0 and all versions and revisions thereafter no longer include this alignment procedure. It utilizes the entered time, date, and location information only to model the sky. Quick Align is an option for equatorial mounts. With an equatorial mount, the alignment can assume a perfect home position on a perfectly polar aligned mount with all the exact time-site settings entered. In reality all of these things vary, but Quick Align can be used as an instant alignment to get the mount in the ballpark, or to simply track and know where the horizons are. With an alt-az mount, such as the NexStar Evolution, the altitude position is usually accounted for (set to the index mark), as is time-site info, but azimuth is a wildcard since the user does not have to point the telescope in any particular position in azimuth, so quick align here does not apply.

7. Last Alignment—Restores the last saved star alignment and switch position. Last Alignment serves as a good safeguard in case the telescope runs out of power.

8. EQ North/EQ South Alignment—When using the optional equatorial wedge and the NexStar Evolution is polar aligned, the EQ North or EQ South alignments offers the choice of performing the Auto Two Star, Two Star, One Star or Solar System alignments.

With the NexStar+ HC plugged into any one of the four AUX port, and the NexStar Evolution is turned on, the display will show "Verifying Packages Please Wait...." After a few moments, the display will show "Evolution" and scroll at the bottom of the screen "Press ENTER to begin alignment". Using the SCROLL buttons on the keypad (numbers 6 and 9), the user can scroll the eight alignment options. To select an option, simply press the ENTER button here the option is displayed.

Following the date screen, the NexStar+ HC will ask the location site of the tele-scope. Accompanying the location site is a scrolling list of cities within the conti-nental United States and foreign countries that can be used so long as the distance from the telescope location is within 50 miles. Or as an option, when the NexStar+ HC asks for a custom site, the latitude and longitude can be entered for greater accuracy. Either way is suitable for the firmware (Figs. 5.2a–l).

Readers should be aware there is a firmware bug with NexStar+ Hand Controls shipped with the early NexStar Evolution telescopes. Firmware version 5.24.4200 contained a program bug that did not allow the resetting of the location site. A workaround of the software bug is to first restore the hand control to its original factory settings, a capability under the Utilities menu. Current (as of February 2016) NexStar+ HC are being shipped with Version 5.28.5200 or newer where the firmware bug has been corrected. Communications between this author and Celestron engineers confirm that if the firmware version of the NexStar+ HC is 5.28.5184 or better, the firmware bug definitely has been corrected. This doesn't mean that version 5.27 or 5.26 does not have the correction. Try it out to see.

If a user owns a NexStar+ HC with version 5.24.4200 or earlier, resetting the location may not be accomplished by accessing Setup Time-Site under Scope Setup. There is a bug with that version of the firmware. Newer firmware is at ver-sion 5.28.xxxx. At the time of this writing, version 5.28.5200 is current. 5.24.4200 was the first version to ship with the Evolution in 2014. Hand controllers with ver-sion 5.28.5184 definitely operate correctly.

A firmware update for the NexStar+ HC is available at: software.Celestron.com/updates/CFM/CFM. This will require an RS-232 cable-to-serial cable and a serial-to-USB connector adapter to connect the NexStar+ HC to a computer. The update is available for both PC and Apple OS. Or as an alternative, contact Celestron and ship the NexStar+ HC to the factory for update.

The best and preferred alignment method for the NexStar Evolution is SkyAlign. The SkyAlign process on the NexStar+ HC is identical as the SkyPortal, select and center on any three bright celestial objects in the sky. Since the NexStar+ HC is not planetarium or graphics based, it is useful to step through the text-based process of SkyAlign.

1. After selecting SkyAlign, the NexStar+ HC will prompt the user to enter the local time using the 24 h format. For example, 5:00 am is 05:00 and 8:30 pm is 20:30. Once the time is entered, hit ENTER.
2. The user is then prompted to select either Daylight Savings Time or Standard Time. Use the Up or Down Scroll keys (number 6 or 9) to toggle between the two selections. When finished, hit ENTER.
3. Select time zone, then enter the date. Hit ENTER each time.
4. After the last ENTER key stroke, SkyAlign will begin. Turn on the red dot finder and using the NexStar+ HC directional arrow buttons, slew the telescope towards

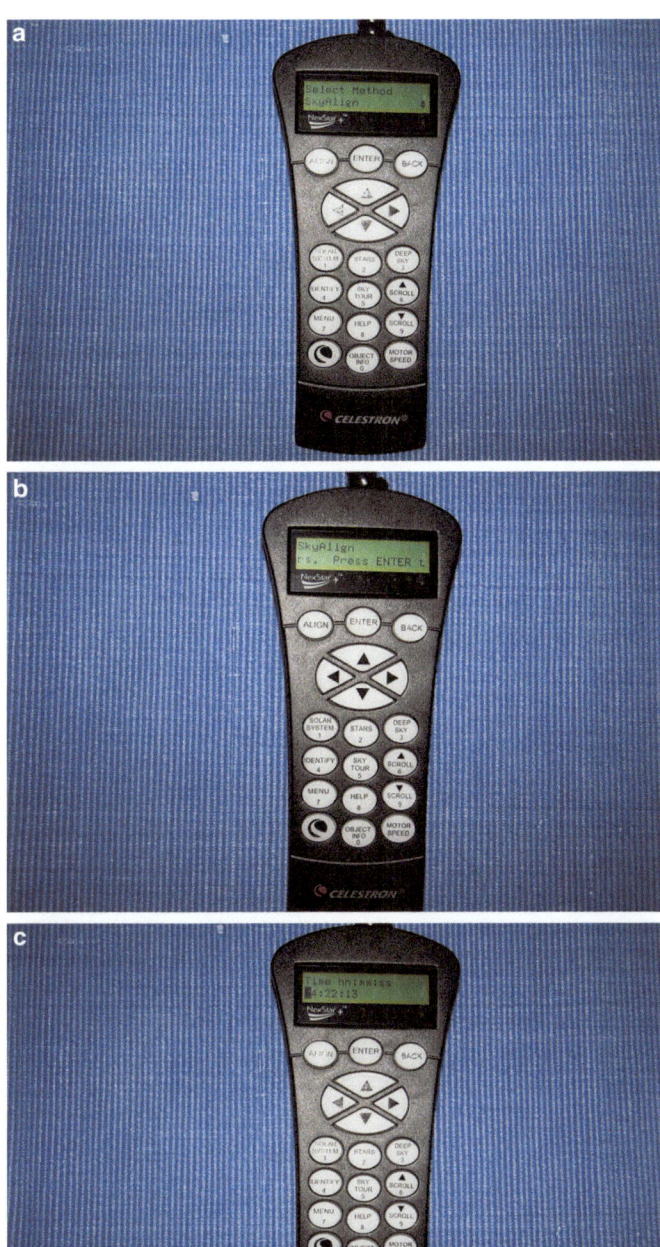

Fig. 5.2 (**a**) Select method alignment screen (Chen) (**b**) SkyAlign screen (Chen) (**c**) Time screen (Chen) (**d**) Daylight savings or standard time screen (Chen) (**e**) Time zone select screen (Chen) (**f**) Date screen (Chen) (**g**) Center object 1 screen (Chen) (**h**) Align object 1 screen (Chen) (**i**) Center object 2 screen (Chen) (**j**) Align object 2 screen (Chen) (**k**) Center object 3 screen (Chen) (**l**) Align object 3 screen (Chen)

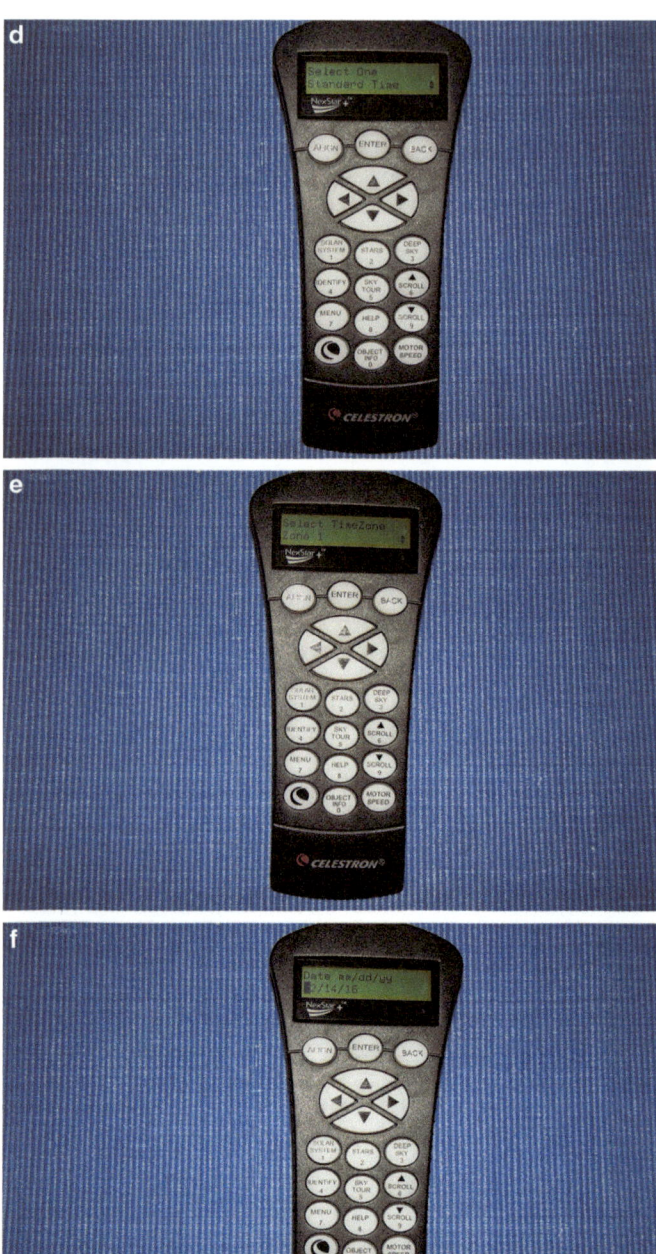

Fig. 5.2 (continued)

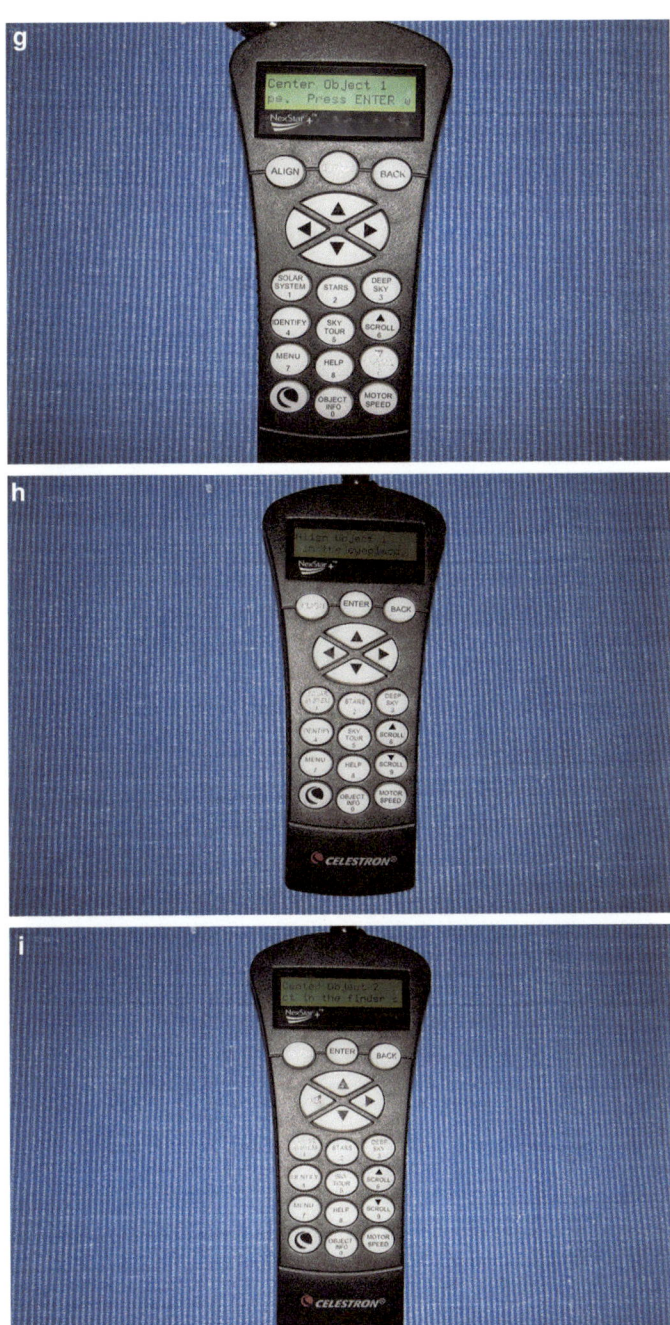

Fig. 5.2 (continued)

Fig. 5.2 (continued)

the first bright star. Center the star by positioning the red dot over the star. Hit ENTER.

5. The NexStar+ HC will then prompt the use the directional arrow buttons, now automatically set to a slower slew rate, to center the bright star in the center of the eyepiece field. Use the Celestron supplied 40 mm Plossl eyepiece for this step. The star should be somewhere in the field-of-view of the eyepiece. Use the directional arrow buttons to center the star. When centered, hit the ALIGN button. The first star will be accepted by the NexStar+ firmware.

6. Select a second bright star at least 20 or 30° away form the first star for the second alignment. Repeat the process described in steps 4 and 5.

7. Select a third bright star, again as far away and at least 20 or 30° from the first and second alignment stars. Repeat the steps 4 and 5 for the third alignment star. When the ALIGN button is pressed, the NexStar+ HC will process the information and if everything was done correctly, the message "Align Success" will appear on the HC display.

When selected, the Auto Two Star alignment is a slightly faster alignment process. SkyAlign provides a more accurate alignment, but the Auto Two Star is a slightly quicker process, where after the first star is selected and centered, the second star will be automatically chosen for the best possible alignment. Once chosen, the NexStar Evolution will automatically slew to the second star to be centered for alignment.

1. After selecting Auto Two Star, the NexStar+ HC will prompt the user to enter the local time using the 24 h format. For example, 5:00 am is 05:00 and 8:30 pm is 20:30. Once the time is entered, hit ENTER.

2. The user is then prompted to select either Daylight Savings Time or Standard Time. Use the Up or Down Scroll keys (number 6 or 9) to toggle between the two selections. When finished, hit ENTER.

3. Select time zone, then enter the date. Hit ENTER each time.

4. After the last ENTER key stroke, Auto Two Star will begin. Turn on the red dot finder and using the NexStar+ HC directional arrow buttons, slew the telescope towards the first bright star. Center the star by positioning the red dot over the star. Hit ENTER.

5. The display will prompt "Select Star 1" from a list of bright stars. Scroll through the list using the "6" and "9" keys. Press ENTER when the desired star appears on the list. Note: whereas SkyAlign does not require any knowledge of star names, this Auto Two Star process does require some knowledge of star nomenclature and their location in the sky.

6. With the red dot finder turned on and the telescope will slew towards the selected bright star. Center the star by positioning the red dot over the star. Hit ENTER.

7. The NexStar+ HC will then prompt the use the directional arrow buttons, now automatically set to a slower slew rate, to center the bright star in the center of the eyepiece field. Use the Celestron supplied 40 mm Plossl eyepiece for this step. The star should be somewhere in the field-of-view of the eyepiece. Use the directional arrow buttons to center the star. When centered, hit the ALIGN button. The first star will be accepted by the NexStar+ firmware.

Fig. 5.3 Auto two star screen (Chen)

8. The NexStar+ hand control will automatically display the most suitable second
   alignment star that is above the horizon. Press ENTER to select and slew towards
   the second star. If a tree, house, or building is blocking the view of the second
   star, press the UNDO button for the next best suitable star or scroll through the
   list of stars using the "6" and "9" keys. Press ENTER when the suitable star is
   displayed.
9. The telescope will slew towards the selected second alignment star. Repeat the
   centering process with the red dot finder and the telescope eyepiece. When the
   ALIGN button is pressed, the NexStar+ HC will process the information and if
   everything was done correctly, the message "Align Success" will appear on the
   HC display (Fig. 5.3).

The One Star Align is a quick-and-dirty alignment procedure using a singular
star for alignment. This method of alignment provides a quick alignment while
sacrificing pointing accuracy. It is suggested that this method be used only under
grab-and-go situations where a limited time is available for alignment and observing, for example a clear sky is available while a bank of clouds is moving in. The
telescope will point in the general area of the searched object, but a low power
eyepiece with a very wide field-of-view will be needed on the telescope.

1. After selecting One Star Align, the NexStar+ HC will prompt the user to enter
   the local time using the 24 h format. For example, 5:00 am is 05:00 and 8:30 pm
   is 20:30. Once the time is entered, hit ENTER.
2. The user is then prompted to select either Daylight Savings Time or Standard
   Time. Use the Up or Down Scroll keys (number 6 or 9) to toggle between the two
   selections. When finished, hit ENTER.

Fig. 5.4 One star align Screen (Chen)

3. Select time zone, then enter the date. Hit ENTER each time.
4. After the last ENTER key stroke, Auto Two Star will begin. Turn on the red dot finder and using the NexStar+ HC directional arrow buttons, slew the telescope towards the first bright star. Center the star by positioning the red dot over the star. Hit ENTER.
5. The display will prompt "Select Star 1" from a list of bright stars. Scroll through the list using the "6" and "9" keys. Press ENTER when the desired star appears on the list. Note: whereas SkyAlign does not require any knowledge of star names, this One Star Align process does require some knowledge of star nomenclature and their location in the sky.
6. With the red dot finder turned on and the telescope will slew towards the selected bright star. Center the star by positioning the red dot over the star. Hit ENTER.
7. The NexStar+ HC will then prompt the use the directional arrow buttons, now automatically set to a slower slew rate, to center the bright star in the center of the eyepiece field. Use the Celestron supplied 40 mm Plossl eyepiece for this step. The star should be somewhere in the field-of-view of the eyepiece. Use the directional arrow buttons to center the star. When centered, hit the ALIGN button. The star will be accepted by the NexStar+ firmware (Fig. 5.4).

The Two Star Align is an alignment procedure using two stars for alignment. This method of alignment provides an alignment procedure that would be suitable under limited sky view conditions, such as viewing from an apartment balcony. This Two Star Align method should provide better GoTo search accuracy than the One Star Align, a low power eyepiece with a very wide field-of-view will still be needed on the telescope for the searched object to fall within the eyepiece field-of-view.

1. After selecting Two Star Align, the NexStar+ HC will prompt the user to enter the local time using the 24 h format. For example, 5:00 am is 05:00 and 8:30 pm is 20:30. Once the time is entered, hit ENTER.

2. The user is then prompted to select either Daylight Savings Time or Standard Time. Use the Up or Down Scroll keys (number 6 or 9) to toggle between the two selections. When finished, hit ENTER.

3. Select time zone, then enter the date. Hit ENTER each time.

4. After the last ENTER key stroke, Auto Two Star will begin. Turn on the red dot finder and using the NexStar+ HC directional arrow buttons, slew the telescope towards the first bright star. Center the star by positioning the red dot over the star. Hit ENTER.

5. The display will prompt "Select Star 1" from a list of bright stars. Scroll through the list using the "6" and "9" keys. Press ENTER when the desired star appears on the list. Note: whereas SkyAlign does not require any knowledge of star names, this Two Star Align process does require some knowledge of star nomenclature and their location in the sky.

6. With the red dot finder turned on and the telescope will slew towards the selected bright star. Center the star by positioning the red dot over the star. Hit ENTER.

7. The NexStar+ HC will then prompt the use the directional arrow buttons, now automatically set to a slower slew rate, to center the bright star in the center of the eyepiece field. Use the Celestron supplied 40 mm Plossl eyepiece for this step. The star should be somewhere in the field-of-view of the eyepiece. Use the directional arrow buttons to center the star. When centered, hit the ALIGN button. The first star will be accepted by the NexStar+ firmware.

8. The display will prompt "Select Star 2" from a list of bright stars. Scroll through the list using the "6" and "9" keys. Press ENTER when the desired star appears on the list.

9. With the red dot finder turned on and the telescope will slew towards the selected bright star. Center the star by positioning the red dot over the star. Hit ENTER.

10. The NexStar+ HC will then prompt the use the directional arrow buttons, now automatically set to a slower slew rate, to center the bright star in the center of the eyepiece field. Use the Celestron supplied 40 mm Plossl eyepiece for this step. The star should be somewhere in the field-of-view of the eyepiece. Use the directional arrow buttons to center the star. When centered, hit the ALIGN button. The second star will be accepted by the NexStar+ firmware and the alignment process is complete (Fig. 5.5).

Solar System Align is a unique alignment procedure designed to provide tracking and GoTo capability during daytime use. The Solar System Align procedure is similar to the One Star Align process detailed earlier. Many observers are not aware that planets and bright stars can be observed in the daytime. The inner planets Mercury and Venus are accessible without having to wait for late afternoon prior to the setting of the Sun or early morning at sunrise. This alignment procedure is used to setup the NexStar Evolution for daytime observation of the Sun with proper solar

Fig. 5.5 Two star align screen (Chen)

filtering. **Please note, any safe observation of the Sun requires the use of proper solar equipment, such as over-the-objective sun filters or solar Hydrogen-alpha filtering. Observing without proper precautions will result in damage to the eyes.**

1. Use the Up and Down buttons (6 and 9 keys) to scroll to Solar Sys.Align and then ENTER to select.
2. The NexStar+ HC will step the user through the time and location process as in previous alignment procedures.
3. Select Object is now displayed. By pressing the Up and Down buttons (6 and 9 keys) selection of the Sun, Moon or planets that are currently between 15 and 70° above the horizon. Note: To include the Sun on the list, first access the Menu button (UNDO back to the "Press ENTER to begin alignment" prompt), select Utilities and then Sun Menu. This is a one-time setting and will be stored for future use. **Please note, any safe observation of the Sun requires the use of proper solar equipment, such as over-the-objective sun filters or solar Hydrogen-alpha filtering. Observing without proper precautions will result in damage to the eyes.**
4. Press ENTER to select the solar system object you wish to use for alignment, use the arrow buttons to slew to the that object. Center the object in the finderscope and press ENTER. Then center the object in the eyepiece and press ALIGN. The Moon works well as an alignment target if it is visible in the daylight. **With proper precautions, using appropriate solar filtering,** the Sun maybe used as an alignment object. Venus is another appropriate alignment target. Locating other planets for alignment during the daytime is somewhat problematic. Experienced observers are able to locate planets during daylight hours, but the

Fig. 5.6 Solar system align screen (Chen)

typical novice may find this to be a problem. Using SkyPortal sky chart (but not as a controller) during the day may provide some aid in locating the somewhat elusive daytime planets.

5. If viewing the object used for alignment is the goal, no additional steps are involved.
6. To provide accurate GoTo and tracking across the sky, use the NexStar+ HC normal GoTo procedures for any object that can be seen in the daylight and use the NexStar Re-Alignment feature to replace the "Unassigned" star with that object (Fig. 5.6).

An optional equatorial wedge is available for the NexStar Evolution, enabling astrophotography and astro-imaging without field rotation. Once the telescope is properly aligned physically with the polar axis of the Earth, EQ North Align or EQ South Align is used on the NexStar+ HC for aligning the NexStar+ firmware for GoTo searches. When tracking equatorially, the NexStar Evolution mount will be moving only on the Right Ascension (RA) axis, with the celestial object centered in the eyepiece field-of-view without rotating.

1. If the user is in the Northern Hemisphere, select EQ North. For those in the Southern Hemisphere, select EQ South. Press ENTER.
2. Step through the time and site process by entering the time, date, and location information as required by the NexStar+ HC.
3. SkyAlign is not available in the EQ procedures. Select EQ AutoAlign. And press ENTER.
4. The telescope will need to be set in a "Home" position. Using the directional arrow keys on the NexStar+ HC, move the telescope until the altitude and meridian markers are aligned. The altitude index marker is located on the top of the fork arm. The meridian marker is located at the base of the fork arm. The tele-

scope tube should end up perpendicular to the fork arm and facing the meridian.

5. The display will prompt "Select Star 1" from a list of bright stars. Scroll through the list using the "6" and "9" keys. Press ENTER when the desired star appears on the list. Note: whereas SkyAlign does not require any knowledge of star names, this Auto Two Star process does require some knowledge of star nomenclature and their location in the sky.

6. With the red dot finder turned on and the telescope will slew towards the selected bright star. Center the star by positioning the red dot over the star. Hit ENTER.

7. The NexStar+ HC will then prompt the use the directional arrow buttons, now automatically set to a slower slew rate, to center the bright star in the center of the eyepiece field. Use the Celestron supplied 40 mm Plossl eyepiece for this step. The star should be somewhere in the field-of-view of the eyepiece. Use the directional arrow buttons to center the star. When centered, hit the ALIGN button. The first star will be accepted by the NexStar+ firmware.

8. The NexStar+ hand control will automatically display the most suitable second alignment star that is above the horizon. Press ENTER to select and slew towards the second star. If a tree, house, or building is blocking the view of the second star, press the UNDO button for the next best suitable star or scroll through the list of stars using the "6" and "9" keys. Press ENTER when the suitable star is displayed.

9. The telescope will slew towards the selected second alignment star. Repeat the centering process with the red dot finder and the telescope eyepiece. When the ALIGN button is pressed, the NexStar+ HC will process the information and if everything was done correctly, the message "Align Success" will appear on the HC display (Figs. 5.7 and 5.8a, b).

Fig. 5.7 Celestron equatorial wedge and NexStar evolution on wedge and tripod (Celestron)

Fig. 5.8 (a) EQNorth align screen (Chen) (b) EQ south align screen (Chen)

NexStar+ Hand Control Settings

There are a number of personal preferences that can be set on the NexStar Evolution telescopes, including the LED brightness of the Celestron logo and WiFi lights, accessory tray lighting, USB charge port and power control.

Press the number 7 key marked "Menu" on the NexStar+ HC. Using the "6" and "9" keys for scrolling up and down the menu list to locate "Peripherals" (Fig. 5.9).

1. Scroll to "Mount Lights" and press ENTER.
2. Scroll to select "Tray" Light and press ENTER. Select a brightness value from 0 (off) to 10 (fully illuminated). Press ENTER.

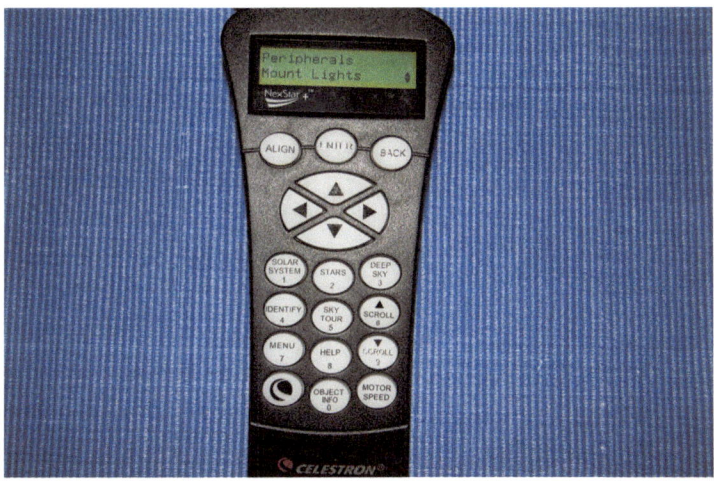

Fig. 5.9 LED brightness adjust (Chen)

3. Scroll to select WiFi Light and press ENTER. Select a brightness value from 0 (off) to 10 (fully illuminated). Press ENTER.
4. Scroll to select Logo and press ENTER. Select a brightness value from 0 (off) to 10 (fully illuminated). Press ENTER.

1. Scroll to "Power" and press ENTER.
2. Scroll to select "Status" or "External Power" and press ENTER. Status will display the battery voltage as High, Medium, or Low. Status also shows charging or discharging. External Power allows the selection of a higher input current from a high output (between 2 and 5 A range) power supply. **Do not use the High setting unless a high output power supply is actually being used**! The standard equipment power supply is a 2.0 amp output (Fig. 5.10).

1. Scroll to "WiFi" and press ENTER.
2. Scroll to select "Status" or "Enable/Disable" and press ENTER. Status displays connection active or inactive, and Direct Connect or Access Point (Fig. 5.11).

1. Scroll to "Charge Port" and press ENTER.
2. Press ENTER to toggle between "Automatic" or "Always On". "Automatic" automatically disables the USB Charge Port when the battery is low. "Always On" keeps the charge port active, even when low battery condition exists (Fig. 5.12).

Fig. 5.10 Power (Chen)

Fig. 5.11 WiFi (Chen)

Fig. 5.12 USB charge port (Chen)

GoTo Searches

GoTo searches are initiated by first pressing the Solar System ("1"), Star ("2"), Deep Sky ("3"), or Tour ("5") keys on the NexStar+ HC (Fig. 5.13).

When the Solar System button is pressed, the NexStar+ HC will display the planets that are visible at the time and location site. The list will be updated as the evening proceeds and additional planets rise and planets set on the horizon. The user scrolls through the list using the Up ("6") and Down ("9") keys, and press ENTER to select and initiate the GoTo search.

Press the BACK key when done with Solar System searches

When the Stars button is pressed, the NexStar+ HC will display the submenu list of star catalogs that are visible at the time and location site.

The user then scrolls the choices of:

1. Constellations
2. Double Stars
3. Named Stars
4. SAO numbered stars—user entry of SAO number
5. Variable stars
6. Asterisms

By scrolling through the listings of catalogs and selecting by pressing ENTER, a list of objects for that catalog becomes available to scroll through. The list will be updated as the evening proceeds and additional objects rise and objects set on the

Fig. 5.13 Solar system search screen (Chen)

horizon. The user scrolls through the list using the Up ("6") and Down ("9") keys, and press ENTER to select and initiate the GoTo search.

Press the BACK key when done with Solar System searches 5.14a-k).

When the Deep Sky button is pressed, the NexStar+ HC will display the submenu list of deep sky objects catalogs that are visible at the time and location site.

The user then scrolls the choices of:

1. Named Objects
2. NGC
3. Abell
4. Caldwell
5. CCD Objects
6. IC
7. Messier

By scrolling through the listings of catalogs and selecting by pressing ENTER, a list of objects for that catalog becomes available to scroll through. The list will be updated as the evening proceeds and additional objects rise and objects set on the horizon. The user scrolls through the list using the Up ("6") and Down ("9") keys, and press ENTER to select and initiate the GoTo search.

Press the BACK key when done with Solar System searches (Fig. 5.15a-b, 5.16a-b, 5.17a-b, 5.18a-b, 5.19a-b, 5.20a-b, 5.21a-b).

Sky Tour is a best of the night, greatest hits sampling of sky for a particular evening. Objects, such as galaxies, globular clusters, open clusters, asterisms, planetary nebulae, diffuse nebulae, planets, and double stars are chosen automatically

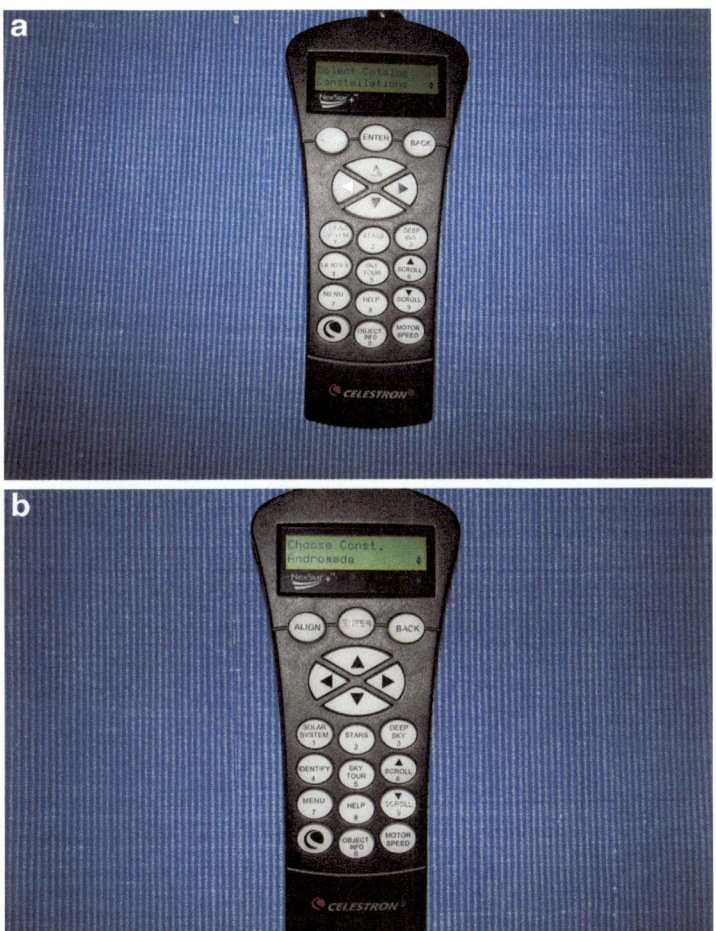

Fig. 5.14 (**a**) Stars search screen, select catalog constellations shown (Chen) (**b**) Constellations select screen (Chen) (**c**) Select catalog double stars screen (Chen) (**d**) Double stars select screen (Chen) (**e**) Select catalog named stars screen (Chen) (**f**) Named star select screen (Chen) (**g**) Select catalog SAO screen (Chen) (**h**) SAO screen for entry of SAO numbers (Chen) (**i**) Select catalog variable stars screen (Chen) (**j**) Variable star select screen (Chen) (**k**) Select catalog asterisms screen (Chen)

Fig. 5.14 (continued)

Fig. 5.14 (continued)

Fig. 5.14 (continued)

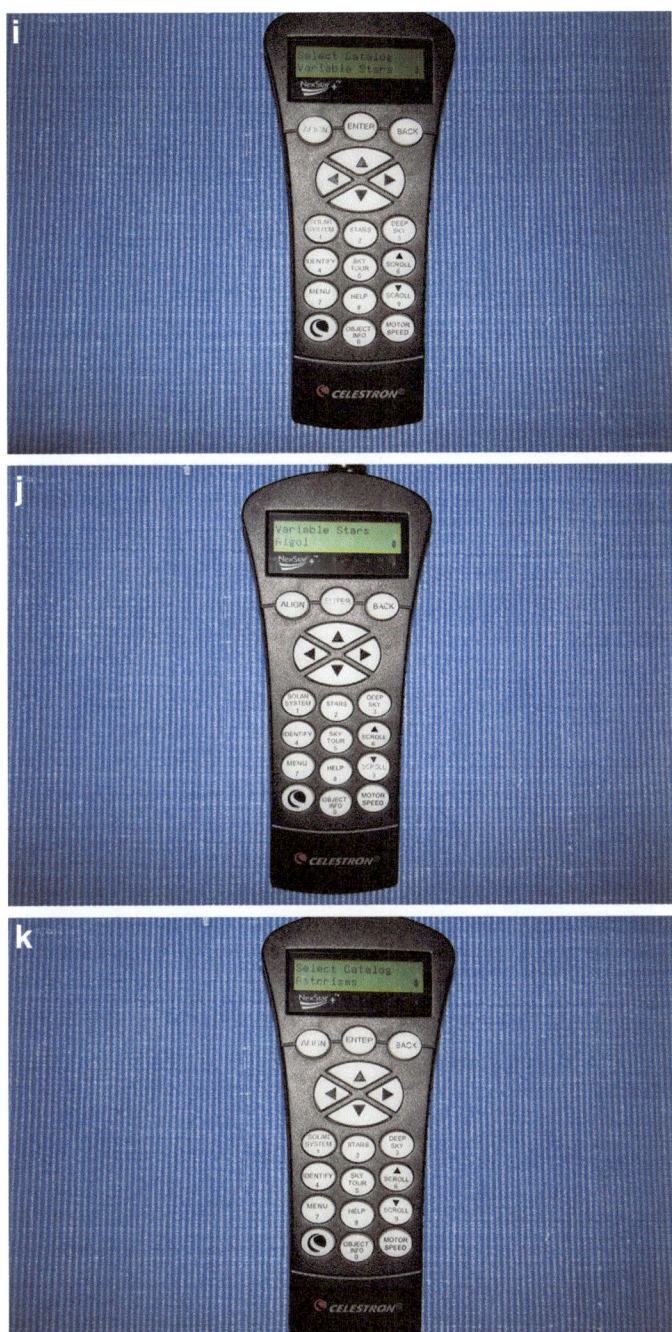

Fig. 5.14 (continued)

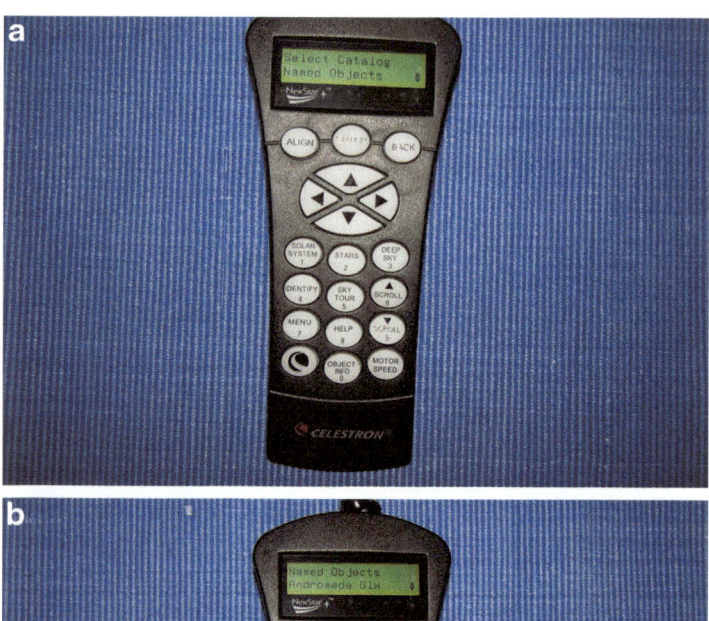

Fig. 5.15 (**a**) Select catalog deep sky named objects screen (Chen) (**b**) Named objects select screen (Chen)

by the NexStar+ firmware. By scrolling through the listings of objects. an object of interest is selected by pressing ENTER. The list is be updated as the evening proceeds and additional objects rise and objects set on the horizon. The user scrolls through the list using the Up ("6") and Down ("9") keys, and press ENTER to select and initiate the GoTo search (Fig. 5.22).

Press the BACK key when done with Sky Tour searches.

Fig. 5.16 (**a**) Select catalog deep sky NGC screen (Chen) (**b**) NGC select screen (Chen)

Utilities

There are a number of utilities that allow the user flexibility in the operation of the NexStar Evolution (Figs. 5.23, 5.24, 5.25, 5.26, 5.27, 5.28, 5.29, 5.30, 5.31 and 5.32).

1. GPS On/Off—This allows the use of Celestron's GPS Module, which is an optional accessory.
2. Factory Setting—This utility returns the telescope back to its factory settings. This is useful when the user is moved the telescope to a different location and/or

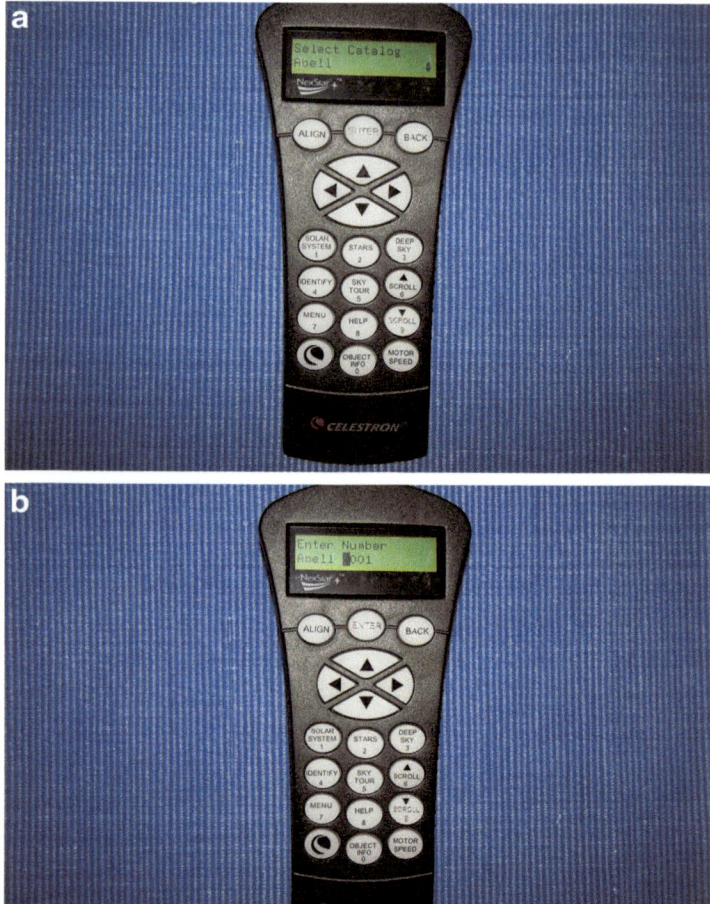

Fig. 5.17 (**a**) Select catalog deep sky Abell screen (Chen) (**b**) Abell select screen (Chen)

time zone. Under normal setups, the NexStar+ HC firmware assumes the is within a reasonable distance from its previous location and will not ask for an update of location site. To reset the location site, the user must use the Factory Setting utility. This will activate the firmware subroutines to ask the user for location site during setup. An alternative to this method is in the Lessons Learned chapter that follows.

3. Version—As with any computer, there is a need to know the current version of the software and, in this case firmware. If necessary, a PC is required to update the firmware in the NexStar+ HC.

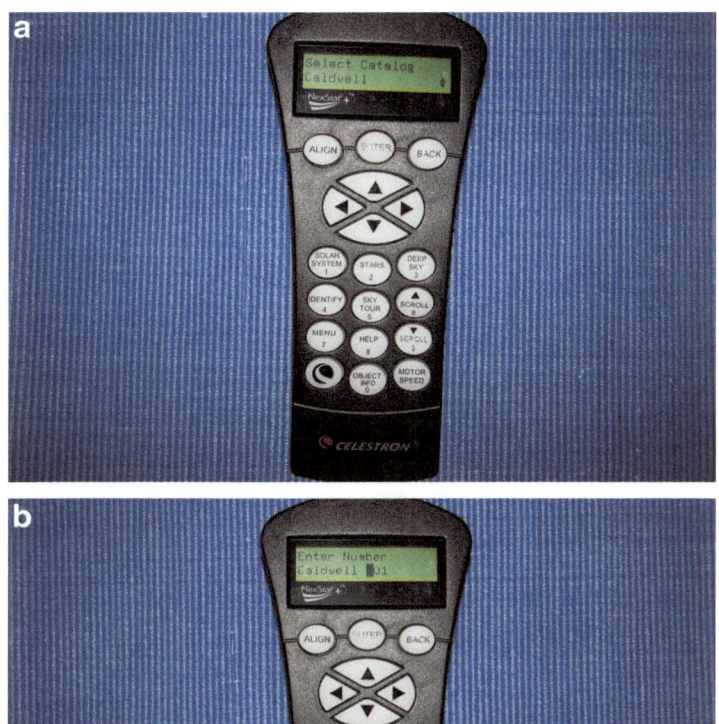

Fig. 5.18 (**a**) Select catalog deep sky Caldwell screen (Chen) (**b**) Caldwell select screen (Chen)

4. Get Axis Position—The altitude and azimuth position of the telescope can be obtained in degrees, minutes, and seconds.
5. GoTo Axis Position—If a specific altitude and azimuth position is required, the altitude and azimuth data can be input here and a GoTo those coordinates can be achieved.
6. Hibernate—If the telescope will not be moved, such as a permanent installation within an observatory, this utility is used to save the GoTo setup and not having to go through the setup process with every use. Access Hibernate in the Utilities menu. The telescope can be positioned anywhere, as long as any change of

Fig. 5.19 (a) Select catalog deep sky CCD objects screen (Chen) (b) CCD objects select screen (Chen)

position is via the hand control. Once Hibernate is enabled, the telescope can be powered off until the next usage. When turning the telescope back on, the user will be prompted to wake the telescope up again. The user will still have to enter the time (since the scope does not have a real time clock built in), but all alignment data saved will be restored. Separately, the SkyPortal app also has a hibernate which works in a similar way, but since all time-site info is in your smart device, the user won't have to enter anything when waking the scope up.

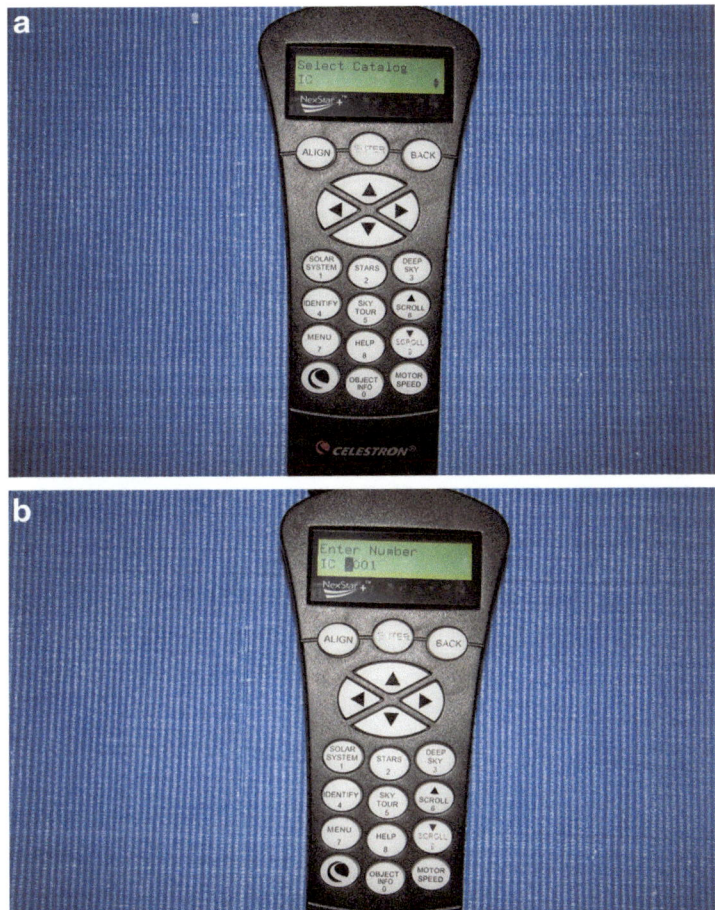

Fig. 5.20 (**a**) Select catalog deep sky IC screen (Chen) (**b**) IC select screen (Chen)

7. Sun Menu—Used during the daytime and allowing the telescope to perform observations of our nearest star. **Please note, any safe observation of the Sun requires the use of proper solar equipment, such as over-the-objective sun filters or solar Hydrogen-alpha filtering. Observing without proper precautions will result in damage to the eyes.**
8. Calibrate GoTo Position—Don't worry about this utility unless a major repair has been performed on the mount.
9. Set Mount Position—Don't worry about this utility. This command will sync the mount to the particular object the telescope is centered on. The usefulness of this command is unknown.

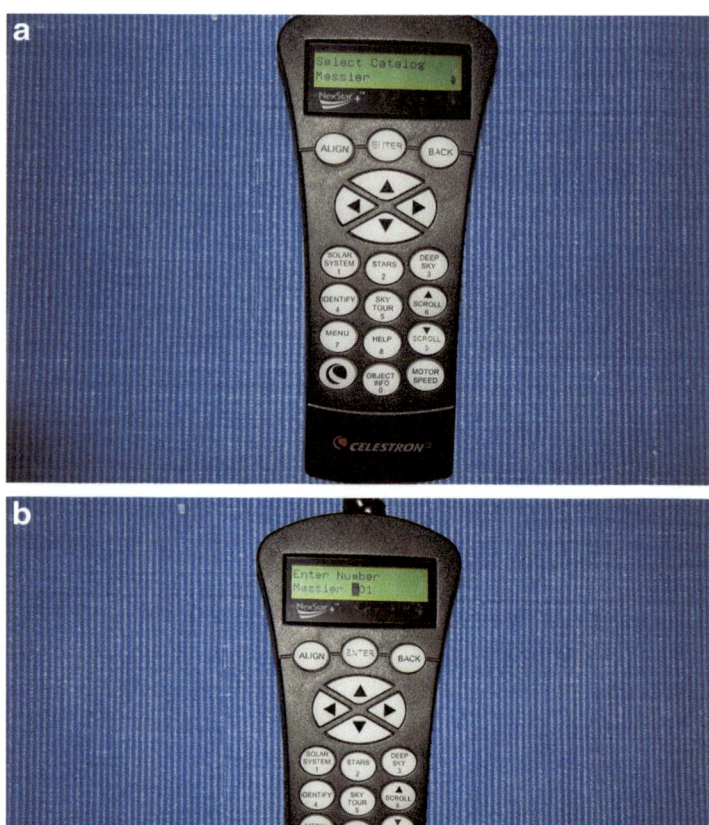

Fig. 5.21 (**a**) Select catalog deep sky messier screen (Chen) (**b**) Messier select screen (Chen)

Fig. 5.22 Sky tour select screen (Chen)

Fig. 5.23 Menu utilities select screen (Chen)

Fig. 5.24 Utilities GPS On/Off screen (Chen)

Fig. 5.25 Utilities factory setting screen (Chen)

Fig. 5.26 Utilities version screen (Chen)

Fig. 5.27 Utilities set axis position screen (Chen)

Fig. 5.28 Utilities GoTo axis position screen (Chen)

Fig. 5.29 Utilities hibernate screen (Chen)

Chapter 6

Lessons Learned in Using the Celestron NexStar Evolution and SkyPortal

Experienced users of computerized GoTo telescopes and mounts are often very adept at jumping and flitting from one telescope to another, one GoTo system to another, without any trouble at all. Just like their high technology cousins the personal computer or the ubiquitous smartphone, there are many commonalities shared among the various systems.

But each GoTo system has its quirks and idiosyncrasies that can prove challenging as the user climbs the learning curve. The SkyPortal presents its own set of challenges, that with user experience, the operation of the system becomes second nature.

Lessons Learned on Using the Celestron NexStar Evolution with SkyPortal

First and foremost, the Evolution telescopes require a good source of electrical power feeding into the system. Whether using the fully charged internal lithium iron-phosphate battery or the Celestron power supply plugged into a wall outlet, the proper voltage and amperage must be supplied to the mount to function properly. A fully charged iPad, iPhone, or Android device with the SkyPortal app is also necessary.

These are general setup tips that apply to the Evolution/SkyPortal mounts:

1. Most NexStar alignment methods do not require the optical tube to be pointed in a specific direction at the beginning of the alignment. There is no "Home" position when the Evolution is used in Alt-Az configuration.

© Springer International Publishing Switzerland 2016   113
J.L. Chen, *The NexStar Evolution and SkyPortal User's Guide*,
The Patrick Moore Practical Astronomy Series,
DOI 10.1007/978-3-319-32539-2_6

2. If using a single alignment point, such as One Star Align or Solar System Align, make sure to level the tripod at the start to get better tracking and more accurate GoTo searches.
3. When using the NexStar+ HC, be accurate to within a couple of minutes when entering the time. Either select a city within 50 miles or enter the location site longitude and latitude to within a degree or two.
4. When using SkyPortal, the latitude, longitude, and local time will be provided by the smart device. Make sure that GPS is enabled on the iOS or Android device. (Non-USA readers are familiar with the term SATNAV in place of the term GPS.)
5. Cell phones and GPS devices are also good sources for accurate time settings when using the Celestron hand controller.
6. The telescope mount tripod should be relatively level. The Evolution telescopes are supplied with a built-in bubble level.
7. Choose alignment stars is different parts of the sky, and at least 10° apart. The Celestron SkyAlign process requires three alignment stars for accurate searches.
8. During GoTo searches, be sure to use a low power wide field eyepiece. A 32–40 mm eyepiece with an apparent field of view of 65° or wider will insure the GoTo search will be successful, with the searched object visible within the field of view. The Evolution series are supplied with a 40 mm low power Plossl eyepiece and a 13 mm medium power Plossl eyepiece as standard equipment.
9. The periodic error correction function is only used for astrophotography. For visual use, this function can be ignored. PEC does not seem to play a part in the search accuracy of the Evolution/SkyPortal system.
10. There is a difference between the iOS version of SkyPortal versus the Android version within the Appearance settings. The iOS SkyPortal Screen Brightness setting is found under Settings > Appearance, with a slider-type screen brightness adjustment. This screen brightness applies only to SkyPortal. For other apps running under iOS, and all Android apps including SkyPortal, the screen brightness must be adjusted from the smart device's main Settings menu.

Prior to setting up your equipment, make sure of the following:

1. Make sure the Evolution internal power source is fully charged. Make sure that the smart device being used for the SkyPortal app is fully charged.
2. If the smart device being used is low in power, remember to bring a USB power cord. Located on the power switch panel at the base of the mount arm, above the two AUX ports is an USB power output port. Connect the smart device here.
3. Disable the sleep or auto-lock mode on the smart device. If the sleep mode is not disabled, the WiFi connection between the Evolution and SkyPortal on the smart device is interrupted when the smart device is inactive too long. In the case of iPhones and iPads, this interruption will result in the device returning to its original WiFi setting (if the home backyard is being used) and causing a complete connect-and-align procedure to be repeated.

4. Close down all apps running in the background on your iOS or Android device. This will free memory for SkyPortal and prevent potential Evolution mount operational oddities.
5. Keep your smart device, smartphone or tablet warm in cold weather. More on this later in this chapter.
6. Make sure the StarPointer red dot finder is properly aligned prior to entering the alignment procedures. Life becomes easier with good finder alignment.
7. During the alignment procedure, lock both axises firmly and never loosen them during your session. All movement of the telescope is done using the electric drive motors of the mount and the Sky Portal controls.
8. Use the manufacturer provided accessories. The back rooms of many telescope stores have examples of telescopes and their mounts with malfunctioning electronics and broken mechanicals because owners have chosen to use off-brand cables, attachments, and the wrong screws. The false economy of a few pennies saved on off-brand accessories will result in tens or hundreds of dollars spent in repairs and hours of lost observing time.
9. Protect the Celestron Evolution and the smart device with the SkyPortal app from the elements and from dropping on the ground. These are a commercial grade electronic products, and not a military-spec or NASA grade device. There are no conformal coatings on the circuit boards to protect them from excessive moisture. The Evolution does not contain shock mount components. Don't abuse the equipment. Take good care of the Celestron and it will stay away from the telescope backroom awaiting repair.
10. If possible, use an inclinometer to set the latitude tilt of the mount when using an equatorial wedge. Here again, the handy smartphone will have an application available to use. Place the smartphone on the right ascension axis of the mount and adjust the tilt of the axis to equal the latitude where you will be observing.

Trouble Shooting the Evolution/SkyPortal System

A number of user problems with the Evolution/Sky Portal have their roots in following the setup data inputs and procedures correctly. The most common symptoms of a setup error occurs when a GoTo command causes the scope to point downward at the Earth, or displaying the dreaded "Object is below the horizon" error message when the GoTo object is clearly above the horizon.

The following is a checklist of possible causes of the downward pointing telescope, inaccurate GoTo slewing, or the dreaded error message:

1. Unlike non-GoTo telescopes, the Celestron StarPointer red dot finder supplied with the Evolution telescopes is a better choice than a upgrade finder scope, such as an 8×50 mm finder. The red dot finder provides a wider viewing angle, correct upright image, and ×1 magnification that won't cause confusion. ×1 magnification will not bring other stars in the view, so the user can zero in on the target bright star, thus yielding a quicker setup time.

2. The use of the Hibernate Enabled is applicable for permanent mounting installations or circumstances where the telescope is not moved. Trying to use the Hibernate Enabled can be tricky and is not recommended for moving back and forth from inside the home to the patio or deck, even if the site is marked. The Hibernate Enabled is really meant for permanent installations of the mount. No axis lock can be loosened when using the Hibernate Enabled. There can be no change of equipment that would change the weight, balance, or configuration of the telescope. Any variation to any of these parameters can cause the telescope to act strangely, such as pointing to the ground or in the opposite direction during a GoTo search.

3. Turn off the Auto-Lock, Sleep, or hibernate settings of the mobile device. When the device goes to sleep because of inactivity, the connection between the SkyPortal app and the Evolution mount is interrupted. The mount will continue to track, but any new searches can't be attempted until the communications between the mobile device and the mount are re-established.

4. The clock and lat/long positioning used by Sky Portal comes from GPS. It is not necessary to account for daylight savings time, the internal GPS of your mobile device will take this into account.

5. During the initial setup, pick three widely spaced alignment stars. GoTo searches will be accurate when this is done, with the searched object falling within the field of view of a low power wide field eyepiece. When SkyAlign is performed properly, a GoTo search will position all of the sought-after objects to appear in the same spot of the eyepiece view. As the observing evening continues, the deep sky objects (especially planetary nebulae) found via GoTo search can then be centered in the eyepiece field and used as additional alignment objects to improve the GoTo accuracy.

6. Choose alignment stars at least 20° above the horizon. The refractive effects of the Earth's atmosphere near the horizon introduces inaccuracies to the alignment.

7. Avoid alignment stars at the zenith. Using straight through finders or red dot finders cause the user to get into an unnatural and uncomfortable position when viewing straight overhead. Proper centering of the alignment star can be compromised, with the user possibly centering the finder on the wrong star.

8. Experience has shown that even with care and following procedures to the letter, sometimes the alignment is off and erroneous GoTo searches occur. Sometimes for unknown reasons the alignment procedure fails, either because of computer glitches or user error. This is when you need to remember TOTOTA—Turn Off, Turn On, Try Again. When glitches occur, switch off the mount and Evolution/Sky Portal, turn it back on and try again. Use a different set of alignment stars the second go around and success should follow.

9. Smart devices are designed for a benign environment. Be aware of the operating and storage specifications of the iPhone and Android smart phones, iPad and Android tablets:

   • Operating temp: 32–95° Fahrenheit
   • Storing temp (turned off): −4 to 113 °F

- Humidity: 5–95 %, non-condensing
- Maximum altitude: 10,000 ft

10. Beware operating the Evolution/Sky Portal in below freezing weather. There have been reports in extreme cold weather that faulty Evolution mounts will not transmit WiFi signals. In this case, contact Celestron for repair. At the lower temperatures, the Apple iOS device or Android device will exhibit a non-responsiveness to commands. The lubricants in the Evolution mount will thicken at extreme low temperatures, and wear and stress on components on the mount becomes a concern. It is recommended that the Evolution mount used in below single digit Fahrenheit temperatures be operated using the Nexstar+ hand controller. Extreme triple digit high Fahrenheit temperatures should also be avoided.

11. Use iOS and Android devices where the ambient temperature is between 0 and 35 °C (32–95 ° F). Low- or high-temperature conditions might cause the device to change its behavior to regulate its temperature. Using these mobile phones or tablets in very cold conditions outside of its operating range may temporarily shorten battery life and could cause the device act strangely, or turn off altogether. Some mobile devices may experience touch-screen inconsistencies, resulting in inaccurate control issues using the directional arrows. Battery life will return to normal when the device is brought back to higher ambient temperatures. In extreme cold conditions, the user is advised to use the NexStar+ hand controller, although it too will experience temperature induced malfunctions in temperature below 0 °F.

If the interior temperature of the mobile device exceeds the normal operating range, the device will protect its internal components by attempting to regulate its temperature. If this occurs, you may notice the following:

1. If the device exceeds a certain temperature threshold, it will present a temperature warning screen similar to this:

- The device stops charging.
- The display dims or goes black

2. In navigation:

- The device will present this alert and turn off the display: "Temperature: iPhone needs to cool down."
- Navigation will continue to provide audible turn-by-turn directions. When approaching a turn, the display will illuminate to guide you through the turn.
- To return the device to normal operation, press the Home Button and slide to unlock. If the device has cooled down enough, you can continue normal usage.

Further Discussion on Cold Weather Operation

Occasionally, during setup or during operation in cold weather, the Evolution mount will exhibit strange behavior. The telescope will "tumble" end-over-end along its altitude axis, the slew speed will be too fast or too slow, or the telescope will not slew at all. SkyPortal controls become unresponsive. These are signs of your mobile device with the SkyPortal app being used at its temperature limits, and not a fault of the Celestron NexStar Evolution or SkyPortal. A too-cold environment is the villain. Here are the alternatives for solving this problem:

1. If using a tablet to operate SkyPortal, switch to a smartphone. After a quick realignment, you'll be back in business. Keep the smartphone in your pocket in-between searches to keep the device warm. If the tablet is the only thing available, place it under the coat if possible to keep it warm and operating properly.
2. Switch to the Celestron NexStar+ hand controller. It will function a little longer in a cold-soaked environment. After a quick realignment, you'll be back in business. Beware that this solution is only temporary, the hand controller can also malfunction eventually in extreme cold temperatures.
3. Or shut the equipment down, bring the Evolution indoors, and go inside and pour yourself a hot chocolate or cup of coffee to warm up. Besides the fact that the equipment doesn't like the cold, you are probably suffering from near-hypothermia too!
4. **DO NOT USE CHEMICAL HAND WARMER PACKS TO KEEP YOUR SMART DEVICE OR HAND CONTROLLER OPERATING**!!! Keep these packs for warming you and not electronics. Their temperatures could raise the temperature of the electronics to the opposite temperature extreme, resulting in electronic failures and damage to your mobile device or hand controller.

Speaking of cold weather, some attention must be paid to making the sure the observer is warm too. Nothing is worse than setting up the NexStar Evolution for a cold nights observing, and then find out that it's too cold for observing despite the equipment working perfectly. Much of this book was written during the late Fall 2015 and Winter 2016, and the research using the NexStar Evolution was conducted during very cold conditions. The historic Blizzard of 2016 yielded 38 in. of snow at the author's home! The NexStar Evolution 6 being used by the author was thoroughly tested in sub-freezing conditions!

Cold Weather Considerations for the Backyard Astronomer

For many backyard astronomers, the fall and winter months are prime time for observing. The skies are dark and transparent because of low humidity. No insects crawling up the legs or flying and buzzing around the head. The trees have lost their leaves and allow more sky to be seen. And rising from his summertime slumber,

Orion rises above the horizon to welcome the astronomers to another wintry observing night.

But boy, is it COLD! There are challenges to amateur astronomy during the winter, all surrounding the fact that it's COLD! The two main challenges are keeping warm, and keeping astronomy equipment functioning.

The following are some common sense suggestions:

1. Check the weather forecast and plan for winter attire as if it's at least 20° colder than the weatherman predicts. Why? This is not a sporting event where a person is in constant motion, and consequently generates internal body heat. Astronomy is an activity that requires lots of sitting or standing at an eyepiece. Without proper precautions, all sort of winter nasties can occur—shivering, frostbite, exposure, and worse. Dress in layers, wear a winter cap or hat, cover the ears, gloves, several layers of socks, boots, long underwear....imagine Ralphie's little brother in the classic movie *A Christmas Story*. By dressing in layers, if it's too warm, just peel off a layer.
2. When it's cold outside, eat a good meal before going out into the field. Also, go to the bathroom before bundling up. Even if the observing site has a rest room or port-a-potty nearby, using the facility can be a challenge at night. Peeling down the layers to "eliminate" is a nuisance and will cause a loss of all the warmth built up.
3. If the winter observing is done in the backyard of your home, take breaks from the eyepiece to go inside and warmup. Just so long as it's dark inside the house to preserve night vision, there is no harm in warming up.
4. Outdoor sports supply stores sell wonderful hand-warmer packs. In fact, Celestron markets a variety of power pack/hand warmers. As previously stated, these hand warmer packs may not be ideal for use in keeping the electronics warm, but they are ideal for keeping hands warm. These work. When hands get cold, these chemically activated packets will do wonders in warming up fingertips.

As the backyard astronomer experiences the cooling weather, some lessons of preparing the telescope equipment are required.

1. Let the telescope and eyepiece acclimate to the ambient temperature. Going from a warm home or car to a cool outdoor environment will require at least 20-to-30 min to adjust to the cooler air.
2. Beware of dewing. As the temperature drops, optics can attract a layer of dew. There are two ways of combating dew: dew shades for the front of the telescope, and dew heater devices to gently maintain the temperature of the optics a few degrees above the dew point. Avoid observing objects directly overhead. Once infected with dew, a brief exposure to the warm air of a hairdryer may help, but then the optics will have to acclimate to the ambient temperature all over again.
3. Beware of fogging, and condensation. It is easy to fog over eyepieces and finderscopes by inadvertently breathing on them. Don't!

4. In early autumn, the bugs and insects that bite and sting may still be a problem. As the weather gets colder, some remaining denizens may find a warmer home in eyepiece cases, telescope cases, and telescopes and mounts. Check all equipment before packing it in for a night.

WiFi Environment

Areas with high WiFi traffic, such as large star parties or crowded public spaces, can potentially pose problems connecting the Evolution telescopes to your mobile smart device. Multiple WiFi telescopes and multiple smart devices can generate conflicting signals.

A practical experiment by the author was conducted by first linking SkyPortal resident on an iPad, then attempting to seize control of the telescope using an iPhone. The result was neither the iPad nor the iPhone were able to control the Evolution 6 on hand. The connection to the iPhone conflicted with the iPad, with the SkyPortal app unable to take control of the telescope.

Celestron offers the following suggestions for remedying these situations for a more reliable WiFi connection to the Evolution or SkyPortal WiFi module:

- If you experience lag or connection problems in areas with high WiFi traffic, try to select a site with minimal WiFi traffic. Numerous nearby devices with WiFi enabled can contribute to unwanted WiFi traffic. Disabling WiFi in nearby devices can improve connectivity to your telescope. The use of the NexStar+ hand controller is advised if the WiFi traffic interferes with the SkyPortal app and your smart device.
- If you have multiple smart devices, and have connected to the same WiFi enabled telescope previously with them, be careful of both devices unintentionally communicating to the telescope at the same time. For example, if you have used both an iPad and iPhone to control your telescope, delete the unused device from the network to prevent conflicts.
- If you experience conflicts when multiple WiFi telescopes are in the same area, such as a star party, and have connected with them with your smartphone or tablet previously, it is recommended to go into settings and delete the other telescopes or devices not currently in use.

NexStar+ Hand Control Reset

A number of users have encountered a NexStar+ hand control oddity in resetting the location site during setup.

It is not uncommon for a NexStar Evolution owner to travel to and setup at a dark site location or star party. Unlike previous generations of NexStar hand controls, resetting the location site on a NexStar+ HC may require the use of the Factory Setting utility.

There is a firmware bug with NexStar+ Hand Controls shipped with the early NexStar Evolution telescopes. Firmware version 5.24.4200 contained a program bug that did not allow the resetting of the location site. A workaround of the software bug is to first restore the hand control to its original factory settings, a capability under the Utilities menu.

The Factory Setting utility returns the telescope back to its factory settings. This is useful when the user is moving the telescope to a different location and/or time zone.

An alternative to this method is to hold down the "0" button on the hand control during power up of the NexStar Evolution mount. This action also results in the firmware to ask the location site, either by closest named city or by latitude and longitude.

Current (as of February, 2016) NexStar+ HC are being shipped with Version 5.28.5200 or newer where the firmware bug has been corrected. Communications between this author and Celestron engineers confirm that if the firmware version of the NexStar+ HC is 5.28.5184 or better, the firmware bug definitely has been corrected. This doesn't mean that version 5.27 or 5.26 does not have the correction. Try it out to see.

If a user owns a NexStar+ HC with version 5.24.4200 or earlier, resetting the location cannot be accomplished by accessing Setup Time-Site under Scope Setup. A firmware update for the NexStar+ HC is available at: software.Celestron.com/updates/CFM/CFM. This will require an RS-232 cable and an RS-232-to-USB connector adapter to connect the NexStar+ HC to a PC computer. Both cables are available through Celestron. Or as an alternative, contact Celestron and ship the NexStar+ HC to the factory for update.

SkyPortal Searches

When scrolling through the various Common Name Lists, but be aware that the lists will often list the same object under different names. For example, under the Messier List, M27 is referred to as the Dumbbell Nebula. Under the Deep Sky Objects list, M27 appears twice, as both the Apple Core Nebula and the more familiar Dumbbell Nebula. Another example is the appearance of M45 as both the Pleiades and the Seven Sisters.

Chapter 7

Introduction to the SkyPortal WiFi Module

The Celestron SkyPortal WiFi module, and its predecessors SkyQLink and SkyQLink2 WiFi modules, are WiFi adapters that allow the alignment and control of non-Evolution Celestron telescopes and equatorial mounts connecting wirelessly using a smartphone or tablet and Celestron's free SkyPortal app.

WiFi Access Using the SkyPortal WiFi Module

The WiFi module is plugged into compatible Celestron mounts via the telescope's Hand Control or AUX port. The module then emits its wireless signal connecting the telescope mount with the iOS or Android smart device. When selecting the WiFi link on the smart device, the WiFi ID of the Evolution or the SkyPortal WiFi module will identify the link as SkyQLink-xx, which is in reference to the original module nomenclature of Celestron. The smart device running the SkyPortal app becomes a wireless hand controller for the non-Evolution telescope or mount, utilizing SkyPortal for the alignment and GoTo searches for any celestial object in the same way as SkyPortal works with Evolution telescopes (Fig. 7.1).

Similar to the operations of the Evolution telescopes, as the SkyPortal app links with a SkyPortal WiFi Module slews the telescope to objects, and provides information in text. The user can also listen to hundreds of included audio descriptions resident on the SkyPortal app on the smart device, explaining the history, mythology, and key features of the most popular celestial objects. Tonight's Best, Messier, Caldwell, Moons, Deep Sky Objects, or any of the others from the search menu and all other SkyPortal functions are available while using the SkyPortal WiFi Module.

© Springer International Publishing Switzerland 2016   123
J.L. Chen, *The NexStar Evolution and SkyPortal User's Guide*,
The Patrick Moore Practical Astronomy Series,
DOI 10.1007/978-3-319-32539-2_7

Fig. 7.1 Celestron WiFi Module (Celestron)

All lists of objects for view are still based on the current exact time and location with the GPS coordinates and the time and date information coming directly from the smart device.

Compatibility Information

SkyPortal WiFi is compatible with the following Celestron mounts:

Advanced VX
CG-5 (requires compatible cable)
CGE Pro
CGEM
CGEM DX
CPC
LCM
NexStar Evolution
NexStar SLT
NexStar SE
SkyProdigy (only works if the automatic alignment is not used)

In the case of the SkyPortal WiFi module being used with the SkyProdigy series of telescopes, the SkyProdigy automatic alignment is incompatible with the workings of the SkyPortal app.

For the WiFi module to work, the automatic alignment procedure must be bypassed.

The StarSense Auto Align accessory is compatible with SkyPortal version 1.5.17 or newer. Prior to December 23, 2015, the StarSense technology was incompatible with SkyPortal, and could only be used with the StarSense hand controller on NexStar Evolution telescopes and other Celestron GoTo mounts. However, with the update SkyPortal Version 1.5.17, this incompatibility was rectified. SkyPortal now supports SkySense AutoAlign. If equipped with the StarSense AutoAlignment accessory, SkyPortal will recognize a StarSense equipped mount through the SkyPortal WiFi module, and the auto-alignment can be accomplished with a single tap of the display screen. The new SkyPortal update supports StarSense EQ, Alt-Az, and Wedge alignment, and StarSense manual align. It is recommended that present users of SkyPortal downloaded prior to December 23, 2015 download the latest version 1.5.17 onto the Apple or Android smart device.

StarSense AutoAlign is now compatible with the following mounts:

1. NexStar Evolution series
2. NexStar SE series
3. NexStar GPS series
4. NexStar SLT with aux splitter
5. Advanced VX
6. CG-5
7. CGEM, CGEM DX
8. CGE
9. CGE Pro
10. CPC, CPC DX
11. LCM with Aux splitter

Please be aware that StarSense AutoAlign is designed for use in the Northern Hemisphere, there may be compatibility issues with equipment used in the Southern Hemisphere.

Chapter 8

Advanced WiFi Tricks

In the previous chapter, some of the lessons learned involved establishing a reliable WiFi connection and involved manipulating the smart device settings.

There are scenarios where the user would want to share multiple devices on one network. For example, if an Evolution/SkyPortal user wants to connect to a home network with internet and control the telescope at the same time. Or still have email access without interfering with the control of the telescope.

This can be done with the Evolution telescopes by setting up in Access Point Mode. The procedure shown here is adapted from the Celestron Evolution manual:

1. Keep the WiFi switch, located on the rear of the fork arm, in the UP position for direct connect.
2. On the smart device settings, connect to the WiFi network "SkyQLink-xx".
3. Open SkyPortal, select Settings, and then tap Telescope Communication.
4. Tap Configure Access Point. Enter the network settings for your network. Enter the exact SSID (or the network broadcast name). Enable DHCP Client, if applicable to the user network, otherwise the user will have to enter the IP Address, Subnet mask, and Gateway for the user's particular network.
5. When complete, select Send Configuration to SkyQLink. A message will appear if SkyQLink was successfully configured.
6. On the telescope, physically move the WiFi switch to the DOWN position. This switches the WiFi to Access Point Mode.
7. Connect to your network with the user's smart device.

© Springer International Publishing Switzerland 2016
J.L. Chen, *The NexStar Evolution and SkyPortal User's Guide*,
The Patrick Moore Practical Astronomy Series,
DOI 10.1007/978-3-319-32539-2_8

8. Open Navigator, select Settings, then select Telescope Communication. tap "Use Access Point". Tap Done when complete.
9. Connect to the telescope in SkyPortal and the connection will now be enabled through the WiFi network.

Once connected to the router the user can then connect the tablet or smartphone to the WiFi network rather than connecting directly to the telescope. For those using the scope in the backyard this might be a better way to use the scope as it might prevent network drop-outs. Also, with the tablet or smartphone connected to the WiFi network instead of the scope itself, the hyperlinks in the SkyPortal software will work.

For those going to star parties, the creation of a local WiFi network may solve the problem of someone else barring the user from connecting to the user's scope without permission. Also there would be no interference as the WiFi would not be advertising itself as open for connections. The answer is a rechargeable portable WiFi router. Seek out one with longer battery life.

There is a scenario that is not possible at the moment. In a classroom situation, where a number of Evolution telescopes are being used by a class of students, it is not possible for the instructor to control all telescopes with one smart device. The limitation is not with the WiFi technology, nor with the Celestron implementation of WiFi. A software modification to SkyPortal, or specifically Sky Safari 4 Basic is required for this multi-telescope capability to happen.

Chapter 9

Accessories for the Celestron NexStar Evolution

This chapter is divided into two categories, one section devoted to accessories to aid (but not limited to) observing and one section devoted to astrophotography and astro-imaging.

Observing Accessories

The lens shade is a flexible plastic shield that attaches to the front of the telescope that has two functions: (1) it aids in the keeping stray light and (2) it shields dew from the front corrector plate of the Celestron SCTs. As a dew shield, the user has to avoid elevating the optical tube towards the zenith for it to be effective. In the presence of dew, the lens shade is most effective in limiting dewing on the front corrector at angles less than 50° above the horizon. The lens shade wraps around the front end of the telescope and is secured to itself by velcro (Fig. 9.1).

Telescope vibration is caused by windy conditions, an unsteady mount or tripod, or even an accidental bump to the instrument, resulting in reduced image quality. Celestron's set of three Vibration Suppression Pads reduces vibration time by almost 100 % and decreases vibration amplitude. The pads fit between the tripod feet and the ground, and is a simple and functional solution to the problem of image disturbance. The pads work on any surface: grass, dirt, concrete, asphalt, wood, etc. (Fig. 9.2).

Celestron Case for NexStar® 4/5/6 & 8″ OTAs, including the Evolution series, is designed to safely transport a NexStar 4/5/6 telescope as well as any Celestron 8″

© Springer International Publishing Switzerland 2016   129
J.L. Chen, *The NexStar Evolution and SkyPortal User's Guide*,
The Patrick Moore Practical Astronomy Series,
DOI 10.1007/978-3-319-32539-2_9

Fig. 9.1 Lens shade (Celestron)

Fig. 9.2 Vibration suppression pads (Celestron)

SCT/EdgeHD optical tube. The case is EVA molded to protect the telescope with a hard yet flexible shell. Also, the case comes with a convenient storage zipper pocket that holds all the telescope's standard accessories and then some. The interior of the case includes two dense foam spacers and a built-in compression strap designed to customize the case for each telescope model. A durable Nylon handle for easy transportation is included (Fig. 9.3).

Fig. 9.3 Celestron case for NexStar 4/5/6 and 8″ optical tube assembly (Celestron)

The Stereo Binocular Viewer allows the user to use two eyepieces and view with both eyes simultaneously. Binocular viewing reduces or eliminates eye fatigue associated with viewing through one eye for a prolonged period of time. This binoviewer produces a three dimensional effect while observing many celestial objects, especially the Moon and the planets. The Stereo Binocular Viewer is designed to work with SCTs without any adaptive optics to compensate for focuser travel (Fig. 9.4).

The Stereo Binocular Viewer requires two eyepieces of equal focal lengths and optical design:

1. High quality FMC BAK-4 prisms to reduce light loss Quality control inspected for collimation and performance.
2. Fully Multi-Coated BAK-4 prisms
3. Foam lined aluminum storage case
4. Eyepiece diopter scales for sharp focus

Fig. 9.4 Stereo binocular viewer (Celestron)

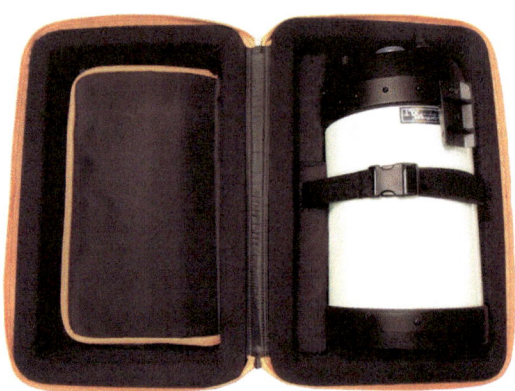

Fig. 9.5 Celestron case for NexStar 8 and 9.25″ & 11″ OTAs (Celestron)

Celestron Case for NexStar 8 and 9.25″ & 11″ OTAs will carry both the Evolution 6 and the Evolution 8 drive base and OTA together. The NexStar carrying case is designed to safely carry the NexStar Evolution 6 or 8 telescope as well as any Celestron 9.25/11″ SCT/EdgeHD optical tube. The case is EVA molded to protect the telescope with a hard yet flexible shell. The case also comes with a convenient storage zipper pocket that holds all the telescope's standard accessories (Fig. 9.5).

The interior of the case includes two dense foam spacers and a built-in compression strap designed to customize the case for each telescope model, and is equipped with durable wheels and an extendable handle for easy transportation.

Fig. 9.6 Mirror diagonal—2″ with XLT coatings (Celestron)

All of the NexStar Evolution telescopes are have f/10 focal ratios, which can limit the field-of-view through 1-1/4 in. eyepieces. Using 2″ eyepieces allows for the use of modern 2″ wide and ultra-wide field designs with apparent FOV of 70 to over 100°. Celestron's offering of a 2″ mirror diagonal with XLT coatings is for all Celestron SCTs, as well as similar competing models and larger Maksutov-Cassegrain designs. This high quality multi-coated mirror diagonal allows the user to take advantage of the wide field of view provided by 2″ eyepieces. Celestron's StarBright XLT coatings are used in its manufacture. This diagonal is interferometer tested to a tenth wave flat mirror (Fig. 9.6).

1. This premium diagonal is designed to work with Schmidt-Cassegrain telescopes.
2. A sturdy aluminum carrying case for travel and protection comes standard.
3. 2″ Diagonal (for SCT)
4. Celestron's high performance XLT coating for maximum reflectivity >96 %
5. Interferometer tested 1/10th wave flat mirror
6. All machined housings for accurate optical alignment
7. Aluminum case for storage and protection

Available in 1-1/4″ and 2″ sizes, The Celestron UHC/LPR filter serves a dual role) (Fig. 9.7).

As a Light Pollution Reduction (LPR) filter, it is designed to selectively reduce the transmission of certain wavelengths of light, specifically those produced by artificial light. This includes mercury vapor, and both high and low pressure sodium vapor lights and the unwanted natural light caused by neutral oxygen emission in our atmosphere (i.e. sky glow).

As a ultra high contrast (UHC) filter, it improves contrast over the typical broadband filters. Sky background is darker, and contrast of emission nebulae are noticeably improved. The advanced technology coatings enable the filter to achieve an outstanding transmission of over 97 % across the entire bandpass,

Fig. 9.7 UHC/LPR light pollution filter (Celestron)

with total blockage of prominent light pollution lines. The perfect filter for viewing nebula from light polluted skies, or for boosting the contrast of nebula from dark sky sites.

In addition to it's optimum spectral and optical characteristics, the UHC/LPR filter offers important features:

1. The multi-layer dielectric coatings are plasma assisted and Ion beam hardened using the latest technology for durability and resistance to scratching.
2. Improved transmission translates to maximum image brightness and contrast. Users of smaller, 4″–11″ telescopes will especially appreciate the high efficiency, and larger scope users will love the rich star fields and detailed subtle nebular shadings that are left intact.
3. The high transmission, sharp cutoffs, and more moderate 60 nm passband of the UHC/LPR filter retains a more natural view, yet significantly boosts overall contrast. Imagers will appreciate the broader bandpass and inclusion of an extremely efficient H-Alpha passband (656 nm).

The 1-1/4″ and 2″ OIII narrowband filter isolates just the two doubly-ionized oxygen lines (496 and 501 nm lines) emitted by planetary and emission nebulae, while blocking the rest of the overall spectrum of light. The result is extreme contrast between the black sky background and the faint photons of OIII light needed for detailed views of the Veil, Ring, Dumbbell, Crescent and Orion nebulae, among other objects (Fig. 9.8).

Each filter has an ultra hard, vacuum-deposited coating carefully designed to block all of the visual spectrum ranging from 400 to 700 nm This eliminates the un-natural colored halos surrounding bright stars common with O III filters of less sophisticated coating technology.

Fig. 9.8 2″ OIII narrowband filter (Celestron)

How well do these filters work? For example, the author used to live in an area with Washington D.C. to the west, Baltimore to the north, and Annapolis MD to the east. To compound the light pollution problems caused by neighborhood street-lights, there was a shopping center ½ mile south and an elementary school 1 mile to the north, both with mercury vapor parking lot lights. Needless to say, the sky was never really dark. In attempting to view the Owl Nebula in Ursa Major, even with a Celestron NexStar GPS 11-in. SCT on a GOTO mount, the nebula could not be seen through the eyepiece without the aid of filtering. Adding a LPR filter darkened the ambient background to allow the Owl Nebula to be seen with averted vision (looking to the side of the object to take advantage of more sensitive areas of the eye). However, replacing the LPR with an O-III filter allowed the direct viewing of the Owl Nebula!

Eyepiece filters are an invaluable aid in lunar and planetary observing. These filters reduce glare and light scattering, increase contrast through selective filtration, increase definition and resolution, reduce irradiation and lessen eye fatigue (Fig. 9.9).

Celestron color filters are mounted in black anodized aluminum cells with the Kodak Wratten Series Numbered individually and engraved, and are available in four assorted kits packed in plastic cases. The cells of each filter are double-threaded, so they can be stacked (piggybacked) in various combinations. This

Fig. 9.9 Celestron eyepiece filter set 1.25 in (21, 80A, 15, 13% neutral density) (Celestron)

allows the user to create different color combinations and transmission characteristics, or to have the same color characteristic, but with a lower transmission. When stacking color filters, the effective transmission of the combination you create is equal to the product of the spectral transmission of each of the filters used.

Celestron's filters are made of high quality, solid plane parallel glass with excellent homogeneity. The filters are anti-reflection coated to prevent glaring and ghosting. All eyepiece filters are threaded to fit Celestron's, and most other manufacturer's, 1¼" eyepieces, and offer a full 26 mm clear aperture.

The Celestron filter set includes the following colors:

– #80A Blue—Enhances the contrast of the cloud belts on Jupiter and the polar ice caps on Saturn. It also can be used to increase contrast on the Moon.
– #21 Orange—Sharpens the boundaries along the plains of Mars due to the reduction of blue/green transmissions. Use on Jupiter and Saturn to enhance detail in the belts and polar regions.
– #12 Deep Yellow—A good general purpose filter for smaller aperture telescopes (less than 4.5″), the #12 Yellow will improve the contrast of features on the Moon. It also aids seeing the equatorial belts on Jupiter and Saturn, clouds and the polar ice caps on Mars, and the dusky detail on Uranus and Neptune.
– ND-96-0.3 Neutral Density—A great filter for the Moon, and for stacking with other planetary filters to increase density without changing the color. A neutral density filter is also useful when splitting binary star systems (double stars). Useful in decreasing brightness and not affecting the color characteristics of an object.

The Equatorial Wedge for NexStar Evolution and SE 6/8 mounts for using the Celestron single fork arm telescopes into portable astro-imaging use (Fig. 9.10).

Fig. 9.10 Equatorial wedge for NexStar evolution

An EQ wedge is necessary to take longer exposure images, or to advance further and use an autoguider.

The normal setup for a NexStar Evolution is in Alt-Az configuration, with the telescope movement left–right and up–down. Alt-Az tracking keeps objects centered, but a long enough exposure will reveal a rotation in the field of view. Adding a wedge allows the mount to track equatorially, thereby eliminating field rotation.

Celestron markets several versions of products under the moniker Elements. These 2-in-1 and 3-in-1 devices combines the features of a hand warmer, flashlight, and a portable power supply for smartphones and personal electronics. Figure 9.11 is an example of one of several versions of these convenient devices that make cold dark night observing more bearable. As hand warmers, these devices warm up to 130°. **Do not use these to warm up smart devices. The 130° temperatures of these as hand warmers exceeds the operating temperature range of all smart devices, and will cause damage iPhones, iPads, and Android devices if used improperly**. The Elements devices can be used to recharge smart phones and tablets.

Worth mentioning are the Celestron PowerTank series of portable batteries for telescope power supplies for field use. Available in 7AH and 17AH sizes, these battery/high power flashlights are mentioned here for academic purposes. Since the NexStar Evolution has its built-in lithium-phosphate battery, the PowerTanks are not needed.

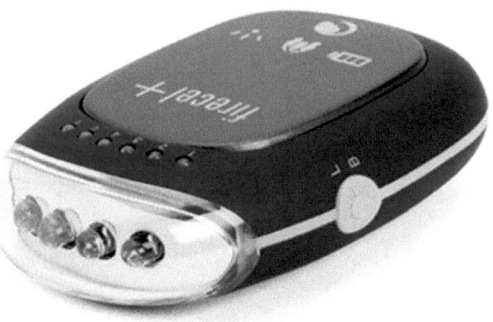

Fig. 9.11 Celestron elements FireCell + (Celestron)

Astrophotography and Astro-Imaging Accessories

A T-Adapter allows the attachment of 35 mm SLR cameras or DSLRs to the prime focus of the telescope. Celestron T-Adapter is for use with all Celestron SCTs. It threads onto the rear cell of the Celestron SCT (Fig. 9.12).

Both a T-Adapter and T-Ring are required to mount a 35 mm SLR camera to the telescope.

The Deluxe Tele-Extender for Schmidt-Cassegrain telescopes is an photographic accessory that allows for eyepiece projection astrophotography or imaging. It consists of a hollow tube that allows the attachment of a camera to the Evolution or any Celestron SCT, with an eyepiece installed. By using the Deluxe Tele-Extender in combination with an eyepiece, there can be an increase the effective focal length to well over 10,000 mm! And this corresponds with a matching increase in image size. The following formula can be used to determine approximate effective focal length when using eyepiece projection photography (Fig. 9.13):

EFL = Telescope focal length/Eyepiece focal length × DF (the distance from the center of the eyepiece to the film).

The Deluxe Tele-Extender is used for high-power lunar, solar and planetary photography as well as for extreme terrestrial photography. It fits over the telescope's eyepiece (even large eyepieces such as those in Celestron's Ultima line), and connects to the visual back of the telescope. Shifting the Evolution tube assembly back or forth on its dovetail will be necessary to properly balance the telescope when using the Tele-Extender.

To use the Deluxe Tele-Extender, remove the diagonal from the telescope and insert an eyepiece directly into the visual back. Place the Tele-Extender over both the eyepiece and the visual back, then attach the 35 mm SLR or DSLR camera to the back of the Tele-Extender, using a T-Ring. The Tele-Extender's built-in safety device will help prevent the eyepiece from becoming accidentally dislodged.

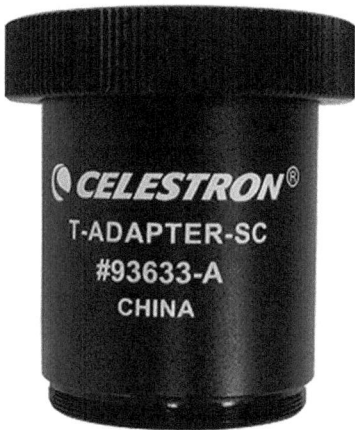

Fig. 9.12 Celestron T-Adapter (Celestron)

Fig. 9.13 Deluxe Tele-Extender for SCTs (Celestron)

There are a few things to keep in mind when using this accessory. First, the image seen through the camera's viewfinder will be upside-down. Secondly, due to the extremely high magnifying effect afforded by this accessory, extra care to prevent camera and telescope vibration. If using the Evolution on an equatorial wedge, accurate polar alignment is required.

NexImage is a one-shot color imager for first-time entry-level astro-imagers that replaces the 1.25″ eyepiece on your telescope and connects to the user's PC via USB 2.0. The 1280×720 CMOS sensor provides high resolution images of the Moon, with enough sensitivity to reveal details on Jupiter and Saturn. NexImage includes Celestron's easy-to-use photo software suite. The iCap capture software and RegiStax stacking software are the same powerful programs included with Celestron's high-end Skyris cameras (Fig. 9.14).

Fig. 9.14 Celestron NexImage solar system imager (Celestron)

With NexImage, the user just points the telescope at the Moon or a planet to record a quick video. The software analyzes each frame of video, throws away the fuzzy ones, and perfectly stacks and aligns the remaining images (for Windows systems only). The result is a bright, clear image with the maximum amount of color and detail. NexImage is a great way to get started with astro-imaging, especially for light-polluted areas.

A/D conversion	10 bit
Mounting	1.25″ barrel
Optical window	IR cutoff filter
Shutter	Electronic rolling shutter
Software compatibility	iCap, IC Capture, DirectShow
Power requirements	Powered by USB
Camera resolution	1280×720
Sensor size	3.86 mm×2.18 mm
Pixel size	3.0 micron square
Frames per second	Up to 30
USB connection	Permanently fixed USB 2.0, 44″ cable length

Skyris is designed and engineered in Germany by The Imaging Source, the leader in high-end planetary imaging, in collaboration with Celestron engineers in California. Powered by SuperSpeed USB 3.0 technology, Skyris transfers data at the fastest possible speeds. Skyris features the Aptina AR0132 CMOS imaging sensor, a state-of-the-art chip used by imaging companies worldwide. Skyris is designed to capture crisp, high-resolution images of the Sun, Moon, and planets.

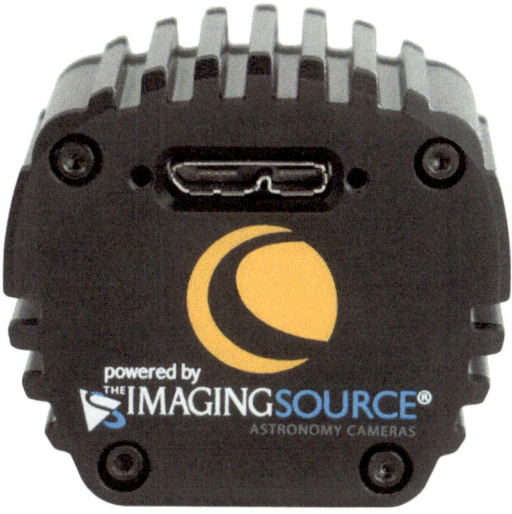

Fig. 9.15 Skyris (Celestron)

This is the step up level in lunar and planetary imaging from the entry-level NexImage (Fig. 9.15).

Skyris is capable of super-fast image download using USB 3.0 connections. The Skyris' CMOS sensor is capable of capturing up to 200 frames per second when sub-framed on planets. Or image the full 1.2 megapixel 1/3-in. sensor at 60 frames per second—great for lunar and solar imaging.

Ultra-Sensitive Aptina AR0132 CMOS Sensor is the CMOS technology that is emerging as the hottest thing in imaging, and Aptina's AR0132 sensor leads the way. This updated version of the popular MT9M034 sensor offers the perfect combination of speed, sensitivity, and value.

Skyris 132C is a one-shot color camera, allowing the capture of full-color images without the need for multiple images using color filters. There's no need to image the same target multiple times to capture different colors, and no investment for color filters and a filter wheel.

The Celestron-designed Skyris camera body helps dissipate heat to minimize the effect of thermal noise on the CCD. Skyris cameras do not have an optical window, which makes cleaning the Skyris camera easier and ensures maximum light transmission with no risk of internal reflections.

All Skyris cameras include Celestron iCap capture software and stacking software (for Windows systems only). This software suite allows the easy control of the camera, capture, and export of images or movie files. Processing includes filtering the images and stacking the best shots to create a high quality image.

The Skyris cameras also make excellent autoguiders. If used with a Celestron mount, Skyris connects as a DirectShow device or WDM-style webcam. Just plug

Fig. 9.16 Universal Piggyback Mount—All SCTs (Celestron)

the hand control into the supplied serial cable using an optional USB to RS-232 Adapter. Skyris works with popular software like MaxIm DL or freeware like PHD Guiding. Or, plug the camera into the ST-4 autoguider port on select non-Celestron mounts, turning Skyris into a high-end autoguider.

The Celestron Piggyback Mount is a great accessory for all observers interested in deep-sky astrophotography. This mount allows the user to attach a camera, with its lens, to the top of the NexStar Evolution Schmidt-Cassegrain telescope (Fig. 9.16).

The Piggyback adapter mounts on top of the rear cell of all Celestron 5, 6, 8, 11, 9.25 and 11-in. Schmidt-Cassegrain telescopes, including the NexStar Evolution series.

Nightscape CCD camera combines the simplicity of a one-shot color imaging camera with the sophisticated features and software of more expensive astronomical imaging systems. Using a Kodak 10.7MP Color Sensor and regulated TEC cooling system, Nightscape can give instant results in just a single exposure. Coupled with the internal mechanical shutter and control software the user can also automatically combine multiple images and dark frames to create images comparable to those taken with professional grade cameras costing thousands more (Fig. 9.17).

Advantages of Nightscape are:

1. Versatility—With compact 4.75 micron×4.75 micron pixels and 2×2 and 4×4 binning, the Nightscape is perfectly matched for optimal resolution at multiple F-numbers. With an F/10 system and 2×2 binning, Nightscape gives you a large image scale while still providing sub-arc second image sampling; ideal for bringing out fine detail in planets and compact deep-sky objects. Or take advantage of the high resolution 4.75 micron×4.75 micron pixels at f/2 to capture favorite

Fig. 9.17 Celestron nightscape CCD camera (Celestron)

wide field objects while maintaining resolution better than existing "seeing" conditions.

2. Fidelity—The Kodak KAI-10100 Color Sensor offers both light sensitivity and a balanced spectral response for capturing true color fidelity in a fraction of the time as with a color filter wheel.

3. Cooling—Regulated Thermoelectric Cooling (TEC) and variable fan control provides a vibration free laminar air flow to dramatically reduce thermal noise inherent in all imaging sensors.

4. Form Factor—Compact 4 in. diameter design provides minimal obstruction needed for Fastar f/2 imaging. The round symmetrical aluminum cast body resembles the obstruction of the secondary mirror of a Cassegrain telescope ensuring round star with minimal light diffraction. Additionally the spacing to sensor is the same distance as most DSLR cameras allowing compatibility with most standard DSLR t-adapters.

5. Full Control—Celestron AstroFX software takes a step-by-step approach from taking images to processing the final result. AstroFX gives full control of the camera from temperature regulation, exposure control as well as computer assisted focusing for easy image acquisition. AstroFX knows just what to do with your images and calibration frames to create a final master image that's been stacked, stretched, sharpened, saturated and ready to share with friends in a snap.

System requirements are limited to PC processors only, with the following requirements:

1. Processor—Pentium™ or equivalent, or higher
2. Windows XP™, Windows Vista™, or Windows 7™ (or higher), 32-bit or 64-bit
3. 1 GB RAM
4. Disk Space—20 MB for program installation
5. Video Display minimum 1024×768, 16-bit color or higher

Unfortunately, Apple OS users need not apply.

Included with the Nightscape is Celestron's innovative AstroFX software. AstroFx takes a step-by-step approach from snapping an image to processing the final result.

AstroFX gives full control of the camera from temperature regulation, exposure control as well as computer assisted focusing for easy image acquisition. AstroFX enables the user to manipulate images and calibrate frames to create a final master image that's been stacked, stretched, sharpened, saturated and ready post on-line or print.

Nightscape CCD Camera—General Features:

1. One shot color imaging using 10.7MP CCD sensor—No filter wheels or color combining needed
2. 4.75 micron×4.75 micron pixels—Small pixels and 2×2 and 4×4 binning allows for optimal resolution at multiple F-numbers
3. Thermoelectric cooling (TEC)—Regulated TEC and adjustable fan to dramatically reduce thermal noise
4. Internal Mechanical Shutter—makes acquiring dark frames easy
5. AstroFX software—takes a step-by-step approach from taking images to processing the final result
6. Full-frame image buffer allows images to be taken even while downloading

Digital Camera Adapter is a universal mounting platform that allows afocal photography (photography through the eyepiece of a telescope) using a 1-1/4″ or 2″ eyepiece. A camera with its camera lens is mounted on the adapter and is aimed at the eyepiece to produce an image. This is a nice entry level device for astrophotography with point-and-shoot cameras (Fig. 9.18).

This combination focal reducer and field corrector lens accessory works with all three NexStar Evolution Schmidt-Cassegrain telescopes. This clever accessory makes it possible to have a dual focal ratio instrument, without sacrificing image quality. The Reducer/Corrector changes the Evolution f/ratio from f/10 to f/6.3, offering a wider field of view. When used for astrophotography, it reduces exposure time by a factor of 3 (Fig. 9.19).

Fig. 9.18 Digital camera adapter, universal (Celestron)

Fig. 9.19 Focal reducer and corrector f/6.3 (Celestron)

Eyepieces

Telescope design is only half of the optical story. The rest of the story are the eyepieces at the focus end of the telescope. There's a marketplace zoo filled with strange and wonderful eyepiece denizens that are vying to complete the optical train. Kellner, Plossl, Abbe, Konig, Brandon, and Nagler are all names of optical designers who have lent their names to their eyepiece designs, and are now an accepted part of the astronomy vocabulary. There are 0.965, 1.25, and 2 in. eyepieces. Some designs have been around for over a century, and some designs are less than a decade old. And most have valid use in astronomy. Since many manufacturers including Celestron offer excellent eyepieces, this discussion is more general in content.

Two of the oldest and simplest of the compound eyepiece designs are the Ramsden and Huygens. Originating from the 1700s, these eyepieces serve as historic curiosities. Occasionally, an antique Ramsden will show up at swap meets, eBay, or even antique stores. The Huygens eyepieces are still supplied in 0.965 in. size on cheap beginner telescopes sold at department stores and big-box stores. Both designs are flawed, with narrow apparent fields, chromatic aberration, and short eye relief (Figs. 9.20 and 9.21).

In the mid-1800s, the Kellner eyepiece was developed by replacing the singlet eye lens element of a Ramsden with a achromat doublet. This resulted in a better performing design with a wider field, better color correction, and less spherical aberrations. When used on long focus telescopes, Kellners still produce a reasonably good image. The main shortcoming of the Kellner is ghosting when looking at bright objects. So Moon watchers beware! The design exists today under various names, including Modified Achromat, RKE, and modified Kellner. These eyepieces are a good economical alternative for those on a budget. For eBay hunters, look for old 1960s–1970s vintage Celestron "volcano top" Kellners for a good value, low cost eyepiece alternatives (Fig. 9.22).

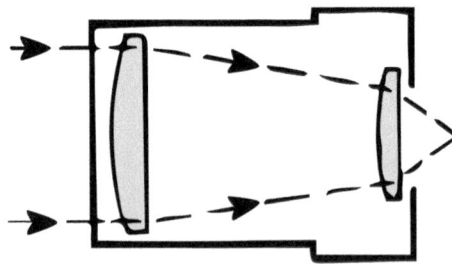

Fig. 9.20 Huygens (Adam Chen)

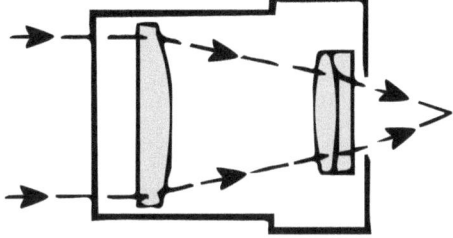

Fig. 9.21 Ramsden (Adam Chen)

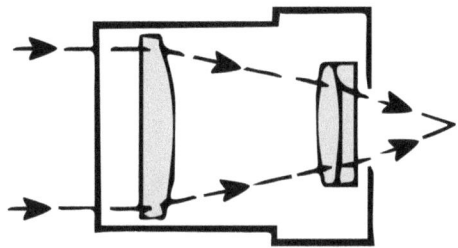

Fig. 9.22 Kellner design (Adam Chen)

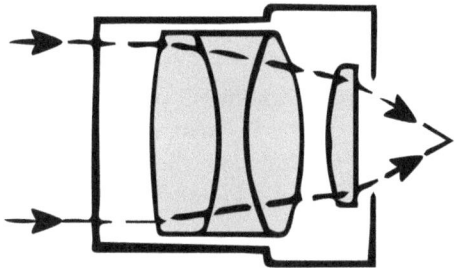

Fig. 9.23 The Abbe orthoscopic (Adam Chen)

An examination of the patents for the Abbe, Plossl, and Konig eyepiece designs describes all three as orthoscopic designs. The term "orthoscopic" means free from distortion. In common astronomy vernacular, the term orthoscopic has evolved to become synonymous in name with the Abbe design (Figs. 9.23 and 9.24).

The Abbe has stood the test of time. Since the 1950s, through the growth of amateur astronomy in the 1960s and 1970s, the Abbe design has been highly regarded for sharp and high contrast images. The classic "volcano top" Abbe ortho-

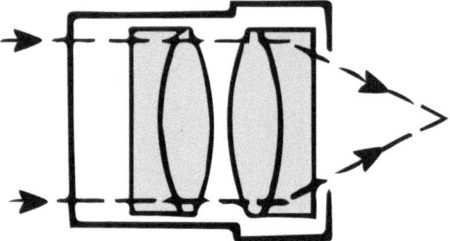

Fig. 9.24 The Plossl orthoscopic (Adam Chen)

scopic, so named for its distinctive beveled shape, are well-known and are highly desired. These Abbe eyepieces have been made by an optician from Japan named Tani-san, whose retirement in 2013 brought an end to decades of Circle T volcano top eyepieces. But don't fret, Abbe orthoscopic eyepieces are available through other sources. The Abbe orthoscopic eyepiece was held up as the pinnacle of eyepieces until the advent of new revolutionary wide-angle and high eye relief eyepieces in the 1980s. Today, dedicated lunar and planetary observers still insist on Abbe eyepieces today. On the used market are examples of Celestron volcano-top Abbe orthoscopics from the 1960s, 1970s and 1980s. They are highly recommended.

The Plossl eyepiece has a interesting reputation in the amateur astronomy world. In the 1960s, the Plossl only existed as a rare and mysterious eyepiece, commercially available from a small vendor in Europe. Then in the 1980s, Plossls suddenly became widely available, to the point now where it is so commonplace that the eyepiece is considered mediocre by average backyard astronomer. Nothing could be further from the truth. A premium Plossl, such as the Celestron Omni series, is an extraordinary eyepiece, versatile in lunar, planetary, and deep sky use. There exists a number of variants of the design in which an extra lens or two to the system, somewhat blurring the definition of a Plossl, but these variants tend to be of high quality and offer high performance.

The NexStar Evolution comes with a 40 and a 13 mm Plossl eyepiece. These are good quality, and can be supplemented with additional eyepieces. Many Celestron customers find the Celestron 1.25″ eyepiece and filter kit sufficiently suits their needs at an economical price point. The kit includes the following:

- Metal, foam-lined carry case
- 32 mm Plossl Eyepiece—1.25″
- 17 mm Plossl Eyepiece—1.25″
- 13 mm Plossl Eyepiece—1.25″ (a duplicate of the Evolution standard equipment)
- 8 mm Plossl Eyepiece—1.25″
- 6 mm Plossl Eyepiece—1.25″
- 2X Barlow Lens—1.25″
- Moon Filter—1.25″

Fig. 9.25 The Konig eyepiece (Adam Chen)

- #80A Blue Filter—1.25″
- #58A Green Filter—1.25″
- #56 Light Green Filter—1.25″
- #25 Red Filter—1.25″
- #21 Orange Filter—1.25″
- #12 Yellow Filter—1.25″

Some dealers, when a customer opts for the eyepiece and filter kit as part of their purchase for the NexStar Evolution, will exchange the duplicate 13 mm Plossl eyepiece for another item. Don't be afraid to ask!

Not as widely available as the other orthoscopics, the Konig design has its fans. Depending on the implementation, the Konig potentially offers a wider field-of-view than the Abbe or Plossl. In practice, the Konig achieves a wider field, but with a slight sacrifice of edge of field sharpness, and slightly shorter eye relief (Fig. 9.25).

A variant of the Konig design is the Brandon eyepiece. Chester Brandon developed his eyepiece design during his time at the Frankford Arsenal in Philadelphia. The eyepiece design was widely used during World War II in U.S. Army optics. The Brandon design differs from the Konig by using three high index glass types and four different lens radii. When marketed in the 1950s as a high priced premium eyepiece, the Brandons were priced at $15.95. My how times have changed, with the price of Brandons and many premium eyepieces in the three-digit range. With sharp crisp contrasty images, the Brandon reputation exceeded that of any other 1950s eyepiece, and is still the go-to eyepiece for many lunar and planetary observers today.

During World War II, a number of scopes used for spotting or for aiming weaponry by the American armed forces were equipped with Erfle eyepieces. In the 1950s and 1960s, war surplus Erfle eyepieces became available for amateur astronomers seeking wide-angle views through their telescopes. With an apparent field of view of approximately 60°, these large eyepieces created a demand for wide field eyepieces. That wide-angle demand has grown over the years, to the point now where the consumer demand for ever wider fields of view drives the eyepiece

a

b

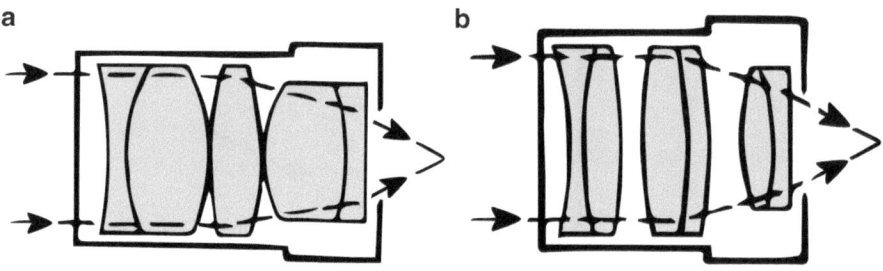

Fig. 9.26 (**a**) The Erfle (Adam Chen). (**b**) Modern wide-angle (Adam Chen)

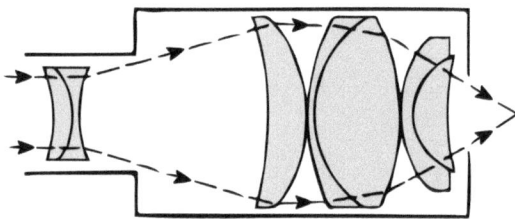

Fig. 9.27 Ultra wide angle eyepieces (Adam Chen)

industry. The Erfle does not display as sharp an image in the center of the field as the orthoscopic eyepiece, and a degradation of the image occurs in the outer third of field. Newer modern wide-field designs, such as the Celestron X-Cel series, use exotic glass types and different lens configuration and curves to correct the edge degradation, while at the same time providing a wider field of view (Fig. 9.26a, b).

The quest for an ever wider field-of-view exploded with the introduction of the Nagler eyepiece. Competitors quickly followed, with virtually every telescope company offering their version of an over 80° field-of-view eyepiece. The Celestron Luminos series offers a range of eyepieces from 7 to 31 mm focal length and 82° AFOV (Fig. 9.27).

The ante was raised again with the introduction 100° and even wider field-of-view eyepieces in recent years. These eyepieces contain seven or more lenses in their complex designs in an effort to provide wide fields without sacrificing sharpness at the edge of the field. These eyepieces are not cheap, with many exceeding the cost of many telescopes! These eyepieces are outstanding for deep sky and wide field applications. However the complex design, high number of optical surfaces, plus the inevitable, although slight, light absorption caused by the amount of glass in the light path, these ultra wide angle eyepieces do not offer the same level of sharpness and contrast as the simpler orthoscopic designs.

While not an eyepiece, a Barlow lens is a useful addition to every eyepiece case. A Barlow lens is a negative lens system placed along the light path between the objective and the eyepiece that increases the effective focal length of the telescope,

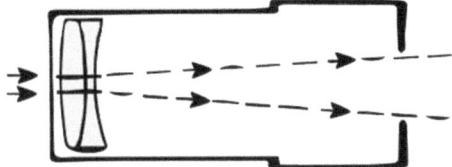

Fig. 9.28 The Barlow lens (Adam Chen)

therefore increasing the magnification. Typically, Barlows double (2x) or triple (3x) the magnification. Newer focal extenders using three or four lens elements are available to quadruple (4x) or quintuple (5x) the focal length. These accessories are useful in three ways. A single Barlow lens effectively doubles the number of magnifications available in an eyepiece collection. The use of a focal extender also allows longer focal length eyepieces with their higher eye relief to be used at higher magnifications for eyeglass wearers. The use of a Barlow lens can improve the off-axis edge sharpness of some eyepiece designs. The Barlow lens and related focal extenders are also useful for astrophotography (Fig. 9.28).

Some discussion is needed on the subject of zoom eyepieces. In the 1960s, the zoom eyepiece earned a reputation for mediocre optics and was not worth the money. Today's zoom eyepieces deserve some attention. Improvements in lens coatings, the introduction of high index glass, and improved manufacturing has yielded a modern zoom that is worthy of a spot in an observer's eyepiece case. Although still narrower in field-of-view at longer focal lengths, and wider at shorter focal lengths, the performance has been greatly improved. A zoom eyepiece will not take the place of an eyepiece collection for critical observing, but serves the role for quick look situations, or when showing the night sky to children whose short attention spans don't allow for the changing and refocusing of conventional eyepieces to change magnification. Celestron offers an excellent 8–24 mm zoom eyepiece.

The driving criteria for eyepiece selection for lunar and planetary observing is sharpness and contrast. The rule of thumb for selecting the right eyepieces for viewing the Moon is "the simpler the better". The classic Abbe, Plossl and Brandon designs are the preferred choices. There are more esoteric lunar and planetary eyepiece designs based on the monocentric design, or on proprietary designs. These are not discussed here due to their low availability.

There is an Achilles heel to the three classic orthoscopic designs. The older eyepiece designs perform best in longer focal length telescopes. In the era in which these designs originated, telescopes had long focal lengths, typically f/10 or greater. Note, all three NexStar Evolution telescopes are f/10 designs, and therefore the classic Plossl and Abbe orthoscopics work well on these telescopes. In today's telescope market, the classic Plossl and Abbe designs are a great deal more affordable than some of the more exotic modern eyepiece designs.

However, for the extended fields of open star clusters, emission and diffuse nebulae, and galaxies, wider field of view would be preferred. Wide angle field of view offered by eyepieces such as Celestron's Ultima Duo or Luminos lines plus a 2″ diagonal will provide excellent performance on the NexStar Evolution telescopes.

Many of today's telescopes, other than the NexStar Evolution series, have much shorter focal lengths, often f/6 or shorter. The classic designs suffer from loss of edge sharpness because of the steeper angle of the light cone from the objective as it enters the eyepiece. Modern designs take into account the shorter focal length telescopes of today. The design rationale for many of the updated configurations of the Plossl design has been to widen the field-of-view and improve performance with short focal length telescopes.

The recommendations for the ideal eyepiece for observing through a NexStar Evolution is as follows:

1. For telescopic views of the Moon and the planets, the Abbe or Plossl designs perform at the highest level. The NexStar Evolution series are f/10 telescopes and these classic eyepiece designs work well.
2. For wide expansive field-of-views of star fields, open clusters, and extended nebulae, the modern wide-angle eyepiece designs are suggested. Eyepieces with apparent field-of-view of 68° or greater provide very appealing views.
3. For observers who must wear glasses while viewing, there are some proprietary eyepiece designs that provide 20 mm of eye relief. Using eyepieces with less eye relief means a loss of field-of-view. Premium eyepieces that use exotic lens configurations and glasses can provide a wide field-of-view with 20 mm of eye relief, but are often not cheap. But they are recommended for eyeglass wearers. The Celestron Ultima Duo series of eyepieces offer excellent 20 mm eye relief that enable eyeglass wearers to observe with their eyeglasses on.
4. Consider using a Barlow in combination with a low- or medium-power eyepiece in order to obtain higher magnifications. The comfortable eye relief from this combination is often preferred by both eyeglass wearers and non-eyeglass wearers. The classic orthoscopic high power (4 and 6 mm) eyepieces are notorious for their near-pinhole sized eye lens. A 12 mm orthoscopic combined with a 2X Barlow yields the same magnification as a 6 mm, but with ample eye relief for eyeglass wearers.

Combination Visual and Astrophotography Accessory

Celestron does offer a unique eyepiece line, with the nomenclature of the Ultima Duo. The Ultima Duo offers performance adaptable to both visual observers and astro-imagers.

The Celestron's Ultima Duo are new eyepieces that combine excellent eyepiece optics with a built-in T-adapter for astro-imaging. With a wide 68° apparent field of view, Ultima Duo provides a generous 20 mm of eye relief and fold-down rubber eye guard make Ultima Duo comfortable to use without removing eyeglasses (Fig. 9.29a).

a

b

Fig. 9.29 (**a**) The Ultima Duo eyepiece (Celestron). (**b**) Telescope smartphone adapters for Ultima Duo eyepieces (Celestron)

For eyepiece projection imaging through the eyepiece, Ultima Duo makes the process quick and easy. A tele-extender is no longer needed. The industry-standard 42 mm T-threads are integrated for a direct, secure connection to a DSLR or astronomical CCD camera. Unscrew and remove the rubber eye guard to expose Ultima Duo's integral T-threads. Then attach the DSLR camera (with compatible T-ring) and astro-imaging can begin. Ultima Duo is ideal for imaging bright solar system objects like the rings of Saturn, craters on the Moon, and Jupiter's Great Red Spot.

Celestron even offers iPhone and Android smartphone users adapters that mount on the Ultima Duo eyepieces for quick-and-easy photos of bright objects, such as the Moon or Jupiter (Fig. 9.29b).

Advanced planetary imagers can attach Celestron's T-to-C Adapter and add a number of high-end imagers to Ultima Duo, including the Skyris camera (Fig. 9.30).

Fig. 9.30 Celestron Ultima Duo T-C adapter (Celestron)

Auto-Alignment

The StarSense Telescope Alignment Accessory from Celestron is an integrated digital camera' that attaches to the telescope's optical tube in place of the finder-scope or red dot finder, and an included hand controller that connects to compatible computerized Celestron telescope mounts. The StarSense technology uses a built-in camera to automatically identify calibration stars and enable go-to alignment of the telescope's optical tube with celestial objects pre-loaded into the StarSense hand controller or SkyPortal app. The camera automatically captures a series of images of the sky. StarSense identifies the stars in the images, matching them to the StarSense hand controller or SkyPortal database. Once a positive match is confirmed, StarSense calculates the coordinates of the center of the captured image, thereby determining exactly where the telescope is pointed. After installing the camera and activating the hand controller, or having SkyPortal auto-detect its presence, the StarSense accessory collects information from its field of view and delivers precise go-to pointing within 3 min (Fig. 9.31).

Prior to December 23, 2015, the StarSense technology was incompatible with SkyPortal, and could only be used with the StarSense hand controller on NexStar Evolution telescopes. However, with the update SkyPortal Version 1.5.17, this incompatibility was rectified. SkyPortal now supports SkySense AutoAlign. If equipped with the StarSense AutoAlignment accessory, a NexStar Evolution and SkyPortal can be aligned with a single tap of the display screen. The new SkyPortal update supports StarSense EQ, Alt-Az, and Wedge alignment, and StarSense manual align. It is recommended that present users of SkyPortal downloaded prior to December 23, 2015 download the latest version 1.5.17 onto the Apple or Android smart device.

Fig. 9.31 StarSense AutoAlign (Celestron)

StarSense AutoAlign is now compatible with the following mounts:

1. NexStar Evolution series
2. NexStar SE series
3. NexStar GPS series
4. NexStar SLT with aux splitter
5. Advanced VX
6. CG-5
7. CGEM, CGEM DX
8. CGE
9. CGE Pro
10. CPC, CPC DX
11. LCM with Aux splitter

Please be aware that StarSense AutoAlign is designed for use in the Northern Hemisphere, there may be compatibility issues with equipment used in the Southern Hemisphere.

Chapter 10

Mounting Other Optical Tubes on the NexStar Evolution Mounts

All Evolution mounts utilize the universal "Vixen-style" dovetail plate system for the quick and easy mounting of telescopes onto the equatorial mount. The system was introduced on the Vixen GP equatorial mounts in 1992. The dovetail system simplified the attachment and removal of telescope optical tubes from mounts, while assuring a firm and safe attachment to the mount when operational. The design has evolved into the most popular of the two standard mounting systems in the amateur astronomy world, with its only rival being the Losmondy Mounting Plate system.

The basis for the mounting system is the use of a basic dovetail, as shown in Fig. 9.1.

The dovetail plate is adaptable to various tube ring or clamshell screw hole configurations of other manufacturers.

Other companies, such as Stellarvue and Orion for example, also provide their telescopes with dovetail plates or rails that attach to their telescopes and fit the standard Vixen Dovetail plate mounting block.

Figure 9.2 provides a sample view of the attachment of a telescope with rings and dovetail plate assembly. Note the position of the tube assembly positioned as far forward as allowed for focuser travel clearance.

The Evolution mount is optimized for use with short tube Celestron 6″, 8″, and 9.25″ SCTs. The mount is easily adaptable to Maksutov–Cassegrain telescopes without the eyepiece diagonal combination bottoming out and striking the Evolution mount base. A short tube 80 or 90 mm f/5 or f/6 refractor fitted with the universal dovetail can be used on the Evolution mount, but care must be taken in properly balancing the OTA on the mount. An additional factor that must be taken into consideration is the focuser and diagonal/eyepiece combination may not have

© Springer International Publishing Switzerland 2016   157
J.L. Chen, *The NexStar Evolution and SkyPortal User's Guide*,
The Patrick Moore Practical Astronomy Series,
DOI 10.1007/978-3-319-32539-2_10

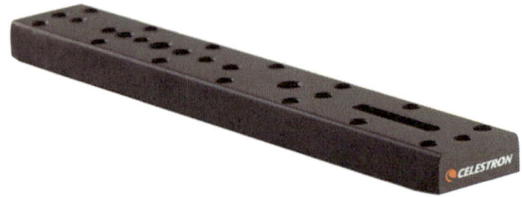

Fig. 9.1 Basic universal dovetail tube plate (Celestron)

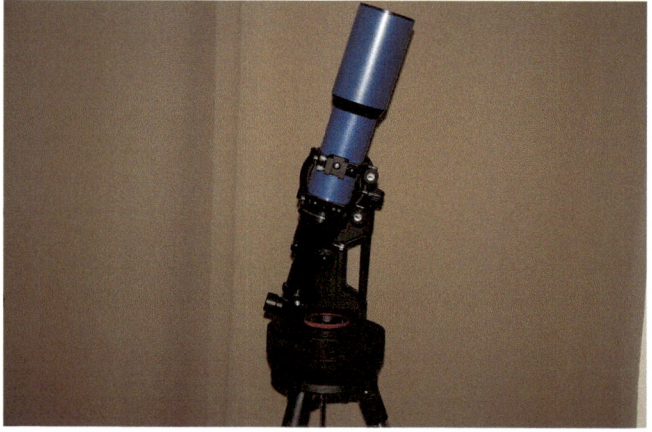

Fig. 9.2 NexStar Evolution mount with an 80 mm refractor mounted (Chen)

the same clearance to safely rotate through the mount vertically, otherwise known as "bottoming out". When faced with this dilemma, SkyPortal allows the user to limit the vertical altitude slewing that will prevent the telescope from striking the mount base. SkyPortal can be adjusted from an angle of 70° above the horizon to a full vertical 90° angle. Access to the full sky maybe limited vertically, but this allows the user some versatility in OTA choice. Note in Fig. 9.3b, with the refractor's focuser tube fully extended, the telescope cannot reach a full vertical position. In this configuration, objects at the zenith cannot be observed.

It is also possible to mount a small short focus Newtonian reflector upon the Evolution mount, with limitations on tube length extending past the dovetail clamp. Depending on the balance of the tube, and how much of the optical tube extends below the dovetail clamp, the mirror end of the Newtonian may or may not interfere with the Evolution mount base. Clearance can be accomplished by either shifting the Newtonian tube forward to achieve clearance, or use the slew limits of SkyPortal to prevent equipment damage.

 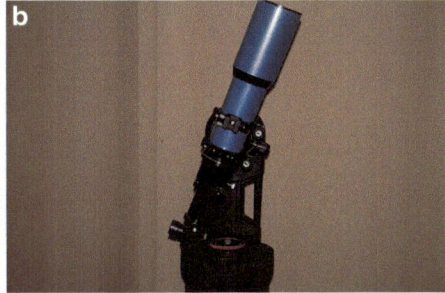

Fig. 9.3 (**a**) NexStar Evolution 6 vertical position diagonal clearance (Chen). (**b**) Evolution Mount with 80 mm refractor vertical position clearance (Chen)

Operating the NexStar Evolution Mount with Another OTA: SkyPortal Versus NexStar+ HC

As stated earlier, when using alternative OTA, the slew limits of SkyPortal must be used to prevent the telescope from hitting the mount.

However, there is no slew limit capability in the NexStar+ HC. Exercise caution if a different (longer tube) telescope is mounted on the NexStar Evolution mount. An errant GoTo search can cause damage to both the OTA and mount when the diagonal and focuser strikes the mount base.

Check for telescope and mount interference, be aware of the limits and if the GoTo slew is headed for disaster, turn off the mount to prevent damage.

The preferred mount control when using an alternate OTA is SkyPortal, with its slew limits. Be very cautious when using the NexStar+HC when an alternate OTA is mounted on the NexStar Evolution mount.

Chapter 11

Maintenance and Care of the NexStar Evolution Mounts and Electronics

Owners of telescopes, telescope mounts, eyepieces, cameras (both digital and film), and numerous accessories have often invested enormous amounts of money in their pursuit of the science of astronomy.

Here is some guidance in the care and feeding of the NexStar Evolution mounts and telescopes.

It is important that owners of NexStar Evolution telescopes recognize that this equipment is made of commercial grade mechanical and electronic components. This is not to criticize Celestron, as the majority of commercially available alt-az and equatorial mounts, and axis drives or GoTo electronics on the market are just that: commercial grade. Components that are MIL-spec or NASA-spec would raise the price of our favorite equipment significantly to the point of not being affordable to the amateur astronomer. Nor do the needs of the average backyard astronomer require MIL-spec or NASA-spec level components.

General Maintenance and Care

The best advice is to treat the NexStar Evolution electronics like you would treat yourself:

Store the Evolution telescopes indoors. An HVAC controlled environment will protect the telescope and mount from damage due to moisture and corrosion. The occasional star party use is okay, so long as the equipment is covered and protected from the weather when not in use.

© Springer International Publishing Switzerland 2016

J.L. Chen, *The NexStar Evolution and SkyPortal User's Guide*,
The Patrick Moore Practical Astronomy Series,
DOI 10.1007/978-3-319-32539-2_11

The appropriateness of a garage or an unheated/non-A/C observatory is debatable. The near-ambient temperature of an unheated garage or observatory means there is no observing time is lost waiting for the telescope optics to adjust to the outdoor ambient temperature. The concern is the presence of moisture and humidity. Corrosion of mechanical and electronic parts can lead to a shortening of the life span of astronomical equipment. A garage environment can be a somewhat better location than an unheated and uninsulated shed. Here again, the reader is reminded that the Evolution electronics and mount are high quality commercial grade products. They are not MIL-spec, and are not built to operate in extreme environments.

Beware of the environmental conditions. There is no published Celestron specification for operating temperature range. The reader is reminded that there is an operating temperature range for the smartphone and tablet of 32–95° F. At the lower temperatures, the smart device will become unresponsive to commands. At extreme low temperatures, pressing the keys will yield no actions by the mount. The lubricant in the Evolution mount will thicken at extreme low temperatures, stressing the drive motors on the mount. It is not recommended that the Evolution be used in single digit Fahrenheit temperatures. Extreme high temperatures should also be avoided. The lubricants of the mount will thin out in extreme high triple digit Fahrenheit temperatures, causing wear on bearings and shortening the mechanical components lifetime. Extreme high triple digit Fahrenheit temperatures have an adverse effect on electronic components, with overheating causing controller failure. Smartphones and tablets will display a warning during high temperature operation, signaling potential damage to electronics and battery due to the high temperature operation. A trip to the repair shop should be avoided at all costs.

Be mindful of the load placed on the Evolution and its mount. A telescope that is overloaded with heavy eyepieces or camera equipment for the mount will exhibit tremors and vibrations that will affect observing, while mechanically stressing the bearings and motor drives. An overloaded and imbalanced mount can shorten the mechanical and electrical life expectancy.

Keep the mount in a clean, dust-free environment. Dirt and dust in mechanical components causes mechanical wear, and a layer of dust on electronic components can cause inadequate cooling, resulting in heat failure of semi-conductors and related components.

Keep the optics clean, but avoid over-cleaning. Follow the cleaning method in the Celestron NexStar Evolution instruction manual.

If a problem arises, consult the Celestron dealer where the mount was purchased. Many problems are covered under warranty. If not covered under warranty, at least the proper repair parts are available.

A Discussion on Dealers, Service, Mail Order, and Warranty

The discussion of warranty coverage leads to a discussion of where to purchase the Celestron NexStar Evolution. If it is geographically convenient, it is highly recommended that the Evolution be purchased in person from an authorized Celestron dealer.

Too many people try to save money by buying on-line, and then expect service from the local dealer. This is a false economy. Believe it or not, the astronomy industry is not a big money, high profit business. Although a dying breed, many telescope retail businesses are Mom and Pop operations run by people who love science and astronomy. They have expertise in amateur astronomy, provide quality products, provide personalized service, and are able to perform many repairs in their own shops. The smaller telescope shops often struggle to compete with high volume Internet or mail-order firms who offer little or no service and rely on manufacturers to repair faulty equipment. Telescope shops, large and small, receive no monetary service support from telescope manufacturers for service, which is very different from automobile dealerships and their service shops. It is unethical to purchase from mail-order and expect a local shop to provide direct service support. Shop at your local telescope store, and support these small business owners.

Consumers need to understand the retail business. There are three criteria for retail competition: Quality, Service and Price. The consumer can only get two of the three. A lower price means the consumer sacrifices either service or quality. If the consumer wants quality and service, be prepared to pay the price. There is no such thing as low price, high quality, and top-notch service. The consumer is best served by purchasing a NexStar Evolution in person from local telescope stores to take advantage of the in-store expertise and in-person service.

SkyPortal Updates

One of the true advantages of controlling a telescope with an application resident on a smartphone or tablet is the ease of updating the application. No longer is it necessary for the user to be computer literate, with an in-depth understanding of port assignments, IP addresses, and the like. When an update is issued, the smart device alerts the user. To update the app, the user can just access the app store and download the latest software update. Gee, that was easy!

Transporting Advice

As seen in the Accessories chapter, Celestron produces carrying cases for the Evolution telescopes. It is always a good idea to protect your investment. A few dollars spent on protective cases for the mount and electronics will pay dividends when the equipment arrives at a remote site or star party intact and is in good working order.

Optics Collimation

The optical alignment of SCTs can occasionally be bumped out of alignment. This is an easy fix. All that is needed is a small Phillips head screwdriver, a calm no wind night, a bright star overhead, and a little patience.

Take the telescope outside and let it come to equilibrium with the ambient air temperature.

After a quick alignment so as to engage the Evolution's tracking, aim the telescope at a bright star near the zenith.

Use a medium to high power eyepiece in the telescope.

Center the bright star in eyepiece and focus in and out. A dark central spot will appear in the out-of-focus image of the star. If the dark central spot, which is caused by the secondary mirror of the SCT design, is in the center and doesn't skewed to one side or the other, the collimation is good and no adjustments are needed.

If the central dark spot is skewed to one side or the other, small adjustments must be made to the secondary mirror to align the optics. Locate the three adjustment screws on the front of the central secondary mirror on the front corrector plate of the telescope.

To adjust the mirror, tighten the screw in small 1/6 to 1/8 turns of the screw toward the direction of the skewed light. The central dark spot should shift towards the center of the out-of-focus star image. Re-centering the star image into the eyepiece FOV maybe necessary to evaluate the adjustment.

If by turning one screw, the other two become loose, then simply tighten the other two screws. Conversely, if the collimation screw gets too tight, then loosen the other two screws by the same amount. The goal is to get the central dark spot in the center of out-of-focus star image and have all screws not loose and relatively tight so as not to move. If too tight, the mirror can flex and distort the image. If the screws are left loose, the mirror could potentially fall out and the secondary could come crashing down on the main mirror, destroying the optics. Be careful!

When the central dark spot is centered, surrounded by concentric rings of light, the task is completed.

If the user is uncomfortable performing this optical alignment task, seek out the local dealer for help. If there is no local dealer, most members of local astronomy clubs can assist in this alignment.

Appendix A

Troubleshooting Checklist

Troubleshooting Steps for the SkyPortal Alignment

If you are having difficulty with your SkyPortal system, please review these steps before calling for technical support:

1. Do you have a fully charged the battery in your mobile device (iPad, iPhone, Android device)? If Yes then proceed, if No, please charge your device's battery before moving forward.
2. Have you disabled the "Sleep" mode on your smart device?
3. Is the tripod relatively level?
4. When Locking the Altitude and Azimuth Axises, be sure they are securely tightened. You may want to try this with the mount off first to see how much pressure is required to lock the axis.
5. Low voltage: Whenever the mount experiences low voltage you may have problems. We strongly encourage all users to fully charge both the mount and smart device prior to the observing session. See step 1!
6. Have you properly aligned the StarPointer red-dot finder?
7. Are you using your lowest power, widest field eyepiece for the alignment procedure?
8. Is the ambient temperature below 32 °F or above 95 °F? If so, take measures to protect your smart device from these cold or heat extremes.

© Springer International Publishing Switzerland 2016
J.L. Chen, *The NexStar Evolution and SkyPortal User's Guide*,
The Patrick Moore Practical Astronomy Series,
DOI 10.1007/978-3-319-32539-2

9. Are the alignment stars in different parts of the sky, and at least 10° apart. The Celestron SkyAlign process requires three alignment stars for accurate searches.

10. Are the alignment stars at least 20° above the horizon? The refractive effects of the Earth's atmosphere near the horizon introduces inaccuracies to the alignment.

11. When all else fails, TOTOTA! Turn Off, Turn On, Try Again!

Appendix B

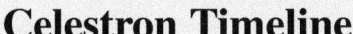

Celestron Timeline

1960   Electronics engineer and owner/president of Valor Electronics Tom Johnson starts an astro-optical division to build telescopes.

1962   The first product is unveiled to the Los Angeles Astronomical Society star party. The 18.5 in. Cassegrain telescope is well-received.

1963   The Celestronic 20 Schmidt-Cassegrain telescope, or C20 is unveiled. Johnson used a novel production method for producing the C20 optics.

1964   The first advertisement for the C20 premieres. By May, the name Celestron Pacific appears in a C20 ad. By December, Valor Electronics disappears, renamed officially Celestron Pacific.

1966   Celestron introduces the first commercially available Schmidt-Cassegrain telescopes (SCTs)

1969   Celestron introduces a full line of SCTs.

1970   Celestron begins printing, supplying, and distributing product catalogs. The orange tube C8 is introduced for $995. 16 in. and 22 in. observatory size telescopes are offered.

1971   Celestron introduces the C-5 5-in. SCT, along with the first commercially available cold camera.

1972   Celestron relocates its production facilities to Torrence, CA. The C-14 14 in. SCT is introduced.

1973   Alan Hale becomes Celestron president.

1976   Celestron branded binoculars are introduced.

1978   Alan Hale becomes CEO of Celestron. New facilities in Torrence CA.

© Springer International Publishing Switzerland 2016   167
J.L. Chen, *The NexStar Evolution and SkyPortal User's Guide*,
The Patrick Moore Practical Astronomy Series,
DOI 10.1007/978-3-319-32539-2

1979   The C-90 Maksutov-Cassegrain is introduced. Spotter or astro versions are availble, with the astro version mounted on a single-arm fork mount with clock drive.

1980   Celestron is sold to a Swiss holding company Diethelm. Johnson retires, and Hale remains   president and CEO of Celestron.

1983   Celestron changes the tube color from orange to black, due to presence of toxic materials in the   orange pigment. Enhanced StarBright coatings are introduced on the new Super C-8.

1984   Celestron partnered with Vixen introduce the first computerized GoTo German equatorial mount, the Super Polaris mount with Sky Sensor.

1987   The author buys his first Celestron telescope, an orange tube C-5.

1996   The Ultima 2000 computerized GoTo 8-in. SCT is introduced.

1997   Celestron introduced the Fastar system, an option on their SCT product line, enabling f/2 imaging with the removal of the Cassegrain secondary and replacing it with a CCD camera designed for this application.

1998   Diethelm sells Celestron to Tasco.

2001   The Nexstar GPS line of GoTo SCTs with integrated GPS is introduced.

2002   Three employees, including Alan Hale, purchase Celestron from Tasco.

2003   Celestron improves its coatings with the introduction of StarBright XLT.

2004   The author acquires a Celestron 11 GPS. ("It followed me home! Can I keep it?")

2005   As a result of a legal dispute with rival Meade Corp., Celestron introduces SkyAlign three star alignment process. Synta Technology Corporation purchases Celestron.

2006   SkyScout Personal Planetarium is introduced. Nexstar SE GoTo SCTs are introduced.

2008   The Celestron CGEM GoTo mount is introduced.

2009   Celestron introduces the EdgeHD line of telescopes and the CGE line of heavy duty GoTo mounts.

2011   StarSense technology is introduced, enabling automatic GoTo alignment using a dedicated CCD imager and on-board firmware.

2014   The Celestron Evolution series of telescopes using Sky Portals Wifi connectivity is introduced.

2016   *The Celestron NexStar Evolution and SkyPortal User Guide* is published.

Appendix C

Celestron NexStar Evolution Schmidt-Cassegrain Telescope Specifications

Celestron NexStar Evolution 6

Optical design	Schmidt-Cassegrain
Aperture	150 mm
Focal length	1500 mm
Focal ratio	10
Focal length of eyepiece 1	40 mm
Focal length of eyepiece 2	13 mm
Magnification of eyepiece 1	×38
Magnification of eyepiece 2	×115
Finderscope	StarPointer
Star diagonal	1.25″
Mount type	Single Fork Arm Altazimuth
Tripod	Stainless steel
Power requirements	Internal LiFePO4 battery included
Highest useful magnification	×354
Lowest useful magnification	×21
Limiting Stellar magnitude	13.4
Resolution (Rayleigh)	0.93 arc s
Resolution (Dawes)	0.77 arc s
Light gathering power (compared to human eye)	×459
Secondary mirror obstruction	2.2 in
Secondary mirror obstruction by area	14 %
Secondary mirror obstruction by diameter	37 %

(continued)

© Springer International Publishing Switzerland 2016

J.L. Chen, *The NexStar Evolution and SkyPortal User's Guide*,
The Patrick Moore Practical Astronomy Series,
DOI 10.1007/978-3-319-32539-2

(continued)

Optical design	Schmidt-Cassegrain
Optical coatings	StarBright XLT
Optical tube length	16 in
Total telescope kit weight	35.4 lbs
Max slew speed	4°/s
Tracking rates	Sidereal, lunar, solar
Alignment procedures	SkyAlign, auto two-star align, two-star align, one-star align, solar system align
Communication ports	One hand control and three aux ports for optional accessories
SkyPortal app database	Over 120,000 including 220 of the best deep sky and solar system objects
USB charge port	Yes

Celestron NexStar Evolution 8

Optical design	Schmidt Cassegrain
Aperture	203.2 mm
Focal length	2032 mm
Focal ratio	10
Focal length of eyepiece 1	40 mm
Focal length of eyepiece 2	13 mm
Magnification of eyepiece 1	×51
Magnification of eyepiece 2	×156
Finderscope	StarPointer
Star diagonal	1.25″
Mount type	Single fork arm altazimuth
Tripod	Stainless steel
Power requirements	Internal LiFePO4 battery included
HIghest useful magnification	×480
Lowest useful magnification	×29
Limiting stellar magnitude	14
Resolution (Rayleigh)	0.69 arc s
Resolution (Dawes)	0.57 arc s
Light gathering power (compared to human eye)	×843
Secondary mirror obstruction	2.5 in. (64 mm)
Secondary mirror obstruction by area	9.77 %
Optical coatings	StarBright XLT
Optical tube length	17 in
Total telescope kit weight	40.6 lbs
Max slew speed	4°/s
Tracking rates	Sidereal, solar and lunar
Alignment procedures	SkyAlign, auto two-star align, two star align, one-star align, solar system align

(continued)

Optical design	Schmidt Cassegrain
Communication ports	One hand control and three aux ports for optional accessories
SkyPortal app database	Over 120,000 including 220 of the best deep sky and solar system objects
USB charge port	Yes

Celestron NexStar Evolution 9.25

Optical design	Schmidt-Cassegrain
Aperture	235 mm
Focal length	2350 mm
Focal ratio	10
Focal length of eyepiece 1	40 mm
Focal length of eyepiece 2	13 mm
Magnification of eyepiece 1	×59
Magnification of eyepiece 2	×180
Finderscope	StarPointer
Star diagonal	1.25″
Mount type	Single fork arm altazimuth
Tripod	Heavy-duty stainless steel
Power requirements	Internal LiFePO4 battery included
Highest useful magnification	×555
Lowest useful magnification	×34
Limiting stellar magnitude	14.4
Resolution (Rayleigh)	0.59 arc s
Resolution (Dawes)	0.49 arc s
Light gathering power (compared to human eye)	×1127
Optical coatings	StarBright XLT
Secondary mirror obstruction	3.35 in (85 mm)
Secondary mirror obstruction by diameter	36 %
Secondary mirror obstruction by area	13 %
Optical tube length	22 in
Total telescope kit weight	46.6 lbs
Max slew speed	4°/s
Tracking rates	Sidereal, solar and lunar
Alignment procedures	SkyAlign, auto two-star align, two star align, one-star align, solar system align
Communication ports	One hand control and three aux ports for optional accessories
SkyPortal app database	Over 120,000 including 220 of the best deep sky and solar system objects
USB charge port	Yes

Appendix D

Messier Catalog

During his lifetime, Charles Messier (1730–1817) was an astronomer noted for his comet discoveries. He found 13 comets and shared in seven more independent co-discoveries. Messier compiled a list of deep sky objects that would were easily confused for comets to help him in his comet searches. Ironically, he is more famous today for his list of non-comet deep sky objects than his comet discoveries. Known as the Messier catalog, it contains 110 objects (actually 109 because of a duplication), including nebulae, clusters and galaxies. The list is fully contained within both the SkyPortal and the Nexstar+ hand controller databases. Many of the Messier objects in the SkyPortal database include photographs of the objects and Celestron Audio.

© Springer International Publishing Switzerland 2016   173
J.L. Chen, *The NexStar Evolution and SkyPortal User's Guide*,
The Patrick Moore Practical Astronomy Series,
DOI 10.1007/978-3-319-32539-2

Messier number	Common name	Constellation	R.A. H:M.S	DEC	App mag	Type
M1	Crab nebula	Taurus	5:34.5	22°01′	8.4	Planetary nebula
M2		Aquarius	21:33.5	−00°49′	6.5	Globular cluster
M3		Canes Venatici	13:42.2	28°23′	6.4	Globular cluster
M4		Scorpius	16:23.6	−26°32′	5.9	Globular cluster
M5		Serpens	15:18.5	2°05′	5.8	Globular cluster
M6	Butterfly cluster	Scorpius	17:40.0	−32°13′	4.2	Open cluster
M7	Ptolemy cluster	Scorpius	17:54.0	−34°49′	3.3	Open cluster
M8	Lagoon nebula	Sagittarius	18:03.7	−24°23′	5.8	Emission nebula
M9		Ophiuchus	17:19.2	−18°31′	7.9	Globular Cluster
M10		Ophiuchus	16:57.2	−4°06′	6.6	Globular cluster
M11	Wild duck cluster	Scutum	18:51.1	−6°16′	5.8	Open cluster
M12		Ophiuchus	16:47.2	−1°57′	6.6	Globular cluster
M13	Hercules cluster	Hercules	16:41.7	36°28′	5.9	Globular cluster
M14		Ophiuchus	17:37.6	−3°15′	7.6	Globular cluster
M15		Pegasus	21:30.0	12°10′	6.4	Globular cluster
M16	Eagle nebula	Serpens	18:18.9	−13°47′	6	Emission nebula
M17	Omega, swan, horseshoe, or lobster nebula	Sagittarius	18:20.8	−16°11′	7	Nebula
M18		Sagittarius	18:19.9	−17°08′	6.9	Open cluster
M19		Ophiuchus	17:02.6	−26°16′	7.2	Globular cluster
M20	Trifid nebula	Sagittarius	18:02.4	−23°02′	8.5	Diffuse nebula
M21		Sagittarius	18:04.7	−22°30′	5.9	Open cluster
M22	Sagittarius cluster	Sagittarius	18:36.4	−23°54′	5.1	Globular cluster
M23		Sagittarius	17:56.9	−19°01′	5.5	Open cluster
M24	Sagittarius star cloud	Sagittarius	18:16.4	−18°29′	4.5	Open cluster
M25		Sagittarius	18:31.7	−19°15′	4.6	Open cluster
M26		Scutum	18:45.2	−9°24′	8	Open cluster
M27	Dumbbell nebula	Vulpecula	19:59.6	22°43′	8.1	Planetary nebula
M28		Sagittarius	18:24.6	−24°52′	6.9	Globular cluster
M29		Cygnus	20:23.0	38°32′	6.6	Open cluster
M30		Capricornus	21:40.4	−23°11′	7.5	Globular cluster
M31	Andromeda galaxy	Andromeda	0:42.7	41°16′	3.4	Spiral galaxy
M32		Andromeda	0:42.7	40°52′	8.2	Elliptical galaxy
M33	Pinwheel galaxy	Triangulum	1:33.8	30°39′	5.7	Spiral galaxy
M34		Perseus	2:42.0	42°47′	5.2	Open cluster

(continued)

(continued)

Messier number	Common name	Constellation	R.A. H:M.S	DEC	App mag	Type
M35		Gemini	6:08.8	24°20′	5.1	Open cluster
M36		Auriga	5:36.3	34°08′	6	Open cluster
M37		Auriga	5:52.0	32°33′	5.6	Open cluster
M38		Auriga	5:28.7	35°50′	6.4	Open cluster
M39		Cygnus	21:32.3	48°26′	4.6	Open cluster
M40	Winnecke 4	Ursa Major	12:22.2	68°05′	8	Dbl star
M41		Canis Major	6:47.0	−20°44′	4.5	Open cluster
M42	Great orion nebula	Orion	5:35.3	−5°27′	4	Nebula
M43	De Mairan's nebula	Orion	5:35.5	−5°16′	9	Nebula
M44	Beehive cluster	Cancer	8:40.0	19°59′	3.1	Open cluster
M45	Pleiades	Taurus	3:47.5	24°07′	1.2	Open cluster
M46		Puppis	7:41.8	−14°49′	6.1	Open cluster
M47		Puppis	7:36.6	−14°30′	4.4	Open cluster
M48		Hydra	8:13.8	−5°48′	5.8	Open cluster
M49		Virgo	12:29.8	8°00′	8.4	Elliptical galaxy
M50		Monoceros	7:03.0	−8°20′	5.9	Open cluster
M51	Whirlpool galaxy	Canes Venatici	13:29.9	47°12′	8.1	Spiral galaxy
M52		Cassiopeia	23:24.2	61°35′	6.9	Open cluster
M53		Coma Berenices	13:12.9	18°10′	7.7	Globular cluster
M54		Sagittarius	18:55.1	−30°29′	7.7	Globular cluster
M55		Sagittarius	19:40 0.0	−30°68′	7	Globular cluster
M56		Lyra	19:16.6	30°11′	8.2	Globular cluster
M57	Ring nebula	Lyra	18:53.6	33°02′	9	Planetary nebula
M68		Virgo	12:37.7	11°49′	9.8	Spiral galaxy
M59		Virgo	12:42.0	11°39′	9.8	Elliptical galaxy
M70		Virgo	12:43.7	11°33′	8.8	Elliptical galaxy
M61		Virgo	12:21.9	4°28′	9.7	Spiral galaxy
M62		Ophiuchus	17:01.2	−30°07′	6.6	Globular cluster
M68	Sunflower galaxy	Canes Venatici	13:15.8	42°02′	8.6	Spiral galaxy
M64	Black eye galaxy	Coma Berenices	12:56.7	21°41′	8.5	Spiral galaxy
M50	Leo's triplet	Leo	11:18.9	13°05′	9.3	Spiral galaxy
M66	Leo's triplet	Leo	11:20.3	12°59′	9	Spiral galaxy
M67		Cancer	8:50.3	11°49′	6.9	Open cluster
M68		Hydra	12:39.5	−26°45′	8.2	Globular cluster
M69		Sagittarius	18:31.4	−32°21′	7.7	Globular cluster
M70		Sagittarius	18:43.2	−32°18′	8.1	Globular cluster
M70		Sagitta	19:53.7	18°47′	8.3	Globular cluster
M72		Aquarius	20:53.5	−12°32′	9.4	Globular cluster
M73		Aquarius	20:68.0	−12°38′		Asterism

(continued)

(continued)

Messier number	Common name	Constellation	R.A. H:M.S	DEC	App mag	Type
M74		Pisces	1:36.7	15°47′	9.2	Spiral galaxy
M75		Sagittarius	20:06.1	−21°55′	8.6	Globular cluster
M76	Cork nebula, little dumbbell	Perseus	1:42.2	51°34′	11.5	Planetary nebula
M77		Cetus	2:42.7	0°01′	8.8	Spiral galaxy
M78		Orion	5:46.7	0°03′	8	Nebula
M79		Lepus	5:24.2	−24°33′	8	Globular cluster
M80		Scorpius	16:17.0	−22°59′	7.2	Globular cluster
M81	Bodes nebula	Ursa Major	9:55.8	69°04′	6.8	Spiral galaxy
M82	Cigar galaxy	Ursa Major	9:56.2	69°41′	8.4	Irregular galaxy
M83	Southern pinwheel galaxy	Hydra	13:37.7	−29°52′	7.6	Spiral galaxy
M84		Virgo	12:25.1	12°53′	9.3	Elliptical galaxy
M85		Coma Berenices	12:25.4	18°11′	9.2	Elliptical galaxy
M86		Virgo	12:26.2	12°57′	9.2	Elliptical galaxy
M87	Virgo A	Virgo	12:30.8	12°24′	8.6	Elliptical galaxy
M88		Coma Berenices	12:32.0	14°25′	9.5	Spiral galaxy
M89		Virgo	12:35.7	12°33′	9.8	Elliptical galaxy
M87		Virgo	12:36.8	13°10′	9.5	Spiral galaxy
M91		Coma Berenices	12:35.4	14°30′	10.2	Spiral galaxy
M92		Hercules	17:17.1	43°08′	6.5	Globular cluster
M93		Puppis	7:44.6	−23°52′	6.2	Open cluster
M94		Canes Venatici	12:50.9	41°07′	8.1	Spiral galaxy
M95		Leo	10:44.0	11°42′	9.7	Spiral galaxy
M96		Leo	10:46.8	11°49′	9.2	Spiral galaxy
M97	Owl nebula	Ursa Major	11:14.9	55°01′	11	Planetary nebula
M98		Coma Berenices	12:13.8	14°54′	10.1	Spiral galaxy
M87	Pin wheel nebula	Coma Berenices	12:18.8	14°25′	9.8	Spiral galaxy
M100		Coma Berenices	12:22.9	15°49′	9.4	Spiral galaxy
M101		Ursa Major	14:03.2	54°21′	7.7	Spiral galaxy
M102	Probably M101 duplicate		14:03.2	54°21′	7.7	duplicate

(continued)

(continued)

Messier number	Common name	Constellation	R.A. H:M.S	DEC	App mag	Type
M103		Cassiopeia	1:33.1	70°42′	7.4	Open cluster
M104	Sombrero galaxy	Virgo	12:40.0	−11°37′	8.3	Spiral galaxy
M105		Leo	10:47.9	12°35′	9.3	Elliptical galaxy
M106		Canes Venatici	12:19.0	47°18′	8.3	Spiral galaxy
M107		Ophiuchus	16:32.5	−13°03′	8.1	Globular cluster
M108		Ursa Major	11:11.6	55°40′	10	Spiral galaxy
M109		Ursa Major	11:57.7	53°23′	9.8	Spiral galaxy
M110		Andromeda	0:40.3	41°41′	8	Elliptical galaxy

Appendix E

The Caldwell Catalog

Sir Patrick Caldwell-Moore, in 1995, noted that the Messier catalog does not include a number of bright deep sky objects, nor does it cover any Southern Hemisphere objects south of declination -35°. He compiled a new catalog to compliment the famous Messier Catalog by including the "missing" objects and to extend the list to cover the Southern Hemisphere.

The resulting list became known as the Caldwell Catalog, which is a collection of 109 of the most impressive celestial objects culled from the NGC and IC catalogs that were not included in Messier's list. Objects in the Caldwell Catalog are organized in descending declination, while objects in Messier's Catalog are listed in order of discovery. The entire Caldwell Catalog is accessible on SkyPortal.

Many of the Caldwell objects in the SkyPortal database include photographs of the objects and Celestron Audio.

© Springer International Publishing Switzerland 2016   179
J.L. Chen, *The NexStar Evolution and SkyPortal User's Guide*,
The Patrick Moore Practical Astronomy Series,
DOI 10.1007/978-3-319-32539-2

C#	NGC/IC	Con.	Type	R. A. h m	Dec. ° ′	Mag.	Size ()	Description
1	188	Cep	OC	00 44.4	+85 20	8.1	14	
2	40	Cep	PN	00 13.0	+72 32	11.6	0.6	
3	4236	Dra	SbG	12 16.7	+69 28	9.7	21×7	
4	7023	Cep	BN	21 01.8	+68 12	6.8	18×18	Reflection nebula
5	IC 342	Cam	SBcG	03 46.8	+68 06	9.2	18×17	
6	6543	Dra	PN	17 58.6	+66 38	8.8	0.3/5.8	Cat's Eye nebula
7	2403	Cam	ScG	07 36.9	+65 36	8.9	18×10	
8	559	Cas	OC	01 29.5	+63 18	9.5	4	
9	Sh2-155	Cep	BN	22 56.8	+62 37	7.7	50×10	Cave nebula
10	663	Cas	OC	01 46.0	+61 15	7.1	16	
11	7635	Cas	BN	23 20.7	+61 12	7.0	15×8	Bubble nebula
12	6946	Cep	ScG	20 34.8	+60 09	9.7	11×9	
13	457	Cas	OC	01 19.1	+58 20	6.4	13	Phi cas cluster
14	869/884	Per	O double C	02 20.0	+57 08	4.3	30 and 30	Sword handle
15	6826	Cyg	PN	19 44.8	+50 31	9.8	0.5/2.3	Blinking nebula
16	7243	Lac	OC	22 15.3	+49 53	6.4	21	
17	147	Cas	dE4G	00 33.2	+48 30	9.3	13×8	
18	185	Cas	dE0G	00 39.0	+48 20	9.2	12×9	
19	IC 5146	Cyg	BN	21 53.5	+47 16	10.0	12×12	Cocoon neb
20	7000	Cyg	BN	20 58.8	+44 20	6.0	120×100	North American nebula
21	4449	CVn	IG	12 28.2	+44 06	9.4	5×3	
22	7662	And	PN	23 25.9	+42 33	9.2	0.3/2.2	
23	891	And	SbG	02 22.6	+42 21	9.9	14×2	
24	1275	Per	Seyfert G	03 19.8	+41 31	11.6	2.6×1	Per A radio source
25	2419	Lyn	GC	07 38.1	+38 53	10.4	4.1	
26	4244	CVn	SG	12 17.5	+37 49	10.6	16×2.5	
27	6888	Cyg	BN	20 12.0	+38 21	7.5	20×10	Crescent nebula
28	752	And	OC	01 57.8	+37 41	5.7	50	
29	5005	CVn	SbG	13 10.9	+37 03	9.8	5.4×2	
30	7331	Peg	SbG	22 37.1	+34 25	9.5	11×4	
31	IC 405	Aur	BN	05 16.2	+34 16	6.0	30×19	Flaming star nebula
32	4631	CVn	ScG	12 42.1	+32 32	9.3	15×3	
33	6992/5	Cyg	SN	20 56.4	+31 43	–	60×8	East veil nebula
34	6960	Cyg	SN	20 45.7	+30 43	–	70×6	West veil nebula
35	4889	Com	E4G	13 00.1	+27 59	11.4	3×2	Brightest in cluster
36	4559	Com	ScG	12 36.0	+27 58	9.8	10×4	
37	6885	Vul	OC	20 12.0	+26 29	5.7	7	
38	4565	Com	SbG	12 36.3	+25 59	9.6	16×3	
39	2392	Gem	PN	07 29.2	+20 55	9.9	0.2/0.7	Eskimo nebula
40	3626	Leo	SbG	11 20.1	+18 21	10.9	3×2	

(continued)

(continued)

C#	NGC/IC	Con.	Type	R. A. h m	Dec. ° ′	Mag.	Size ()	Description
41	–	Tau	OC	04 27.0	+16 00	1.0	330	Hyades
42	7006	Del	GC	21 01.5	+16 11	10.6	2.8	Very distant globular
43	7814	Peg	SbG	00 03.3	+16 09	10.5	6×2	
44	7479	Peg	SBbG	23 04.9	+12 19	11.0	4×3	
45	5248	Boo	ScG	13 37.5	+08 53	10.2	6×4	
46	2261	Mon	BN	06 39.2	+08 44	10.0	2×1	Hubble's variable neb.
47	6934	Del	GC	20 34.2	+07 24	8.9	5.9	
48	2775	Can	SaG	09 10.3	+07 02	10.3	4.5×3	
49	2237 -9	Mon	BN	06 32.3	+05 03	–	80×60	Rosette nebula
50	2244	Mon	OC	06 32.4	+04 52	4.8	24	
51	IC 1613	Cet	IG	01 04.8	+02 07	9.0	12×11	
52	4697	Vir	E4G	12 48.6	−05 48	9.3	6×3	
53	3115	Sex	E6G	10 05.2	−07 43	9.1	8×3	Spindle galaxy
54	2506	Mon	OC	08 00.2	−10 47	7.6	7	
55	7009	Aqr	PN	21 04.2	−11 22	8.3	2.5/1	Saturn nebula
56	246	Cet	PN	00 47.0	−11 53	8.0	3.8	
57	6822	Sgr	IG	19 44.9	−14 48	9.3	10×9	Barnard's galaxy
58	2360	CMa	OC	07 17.8	−15 37	7.2	13	
59	3242	Hya	PN	10 24.8	−18 38	8.6	0.3/21	Ghost of jupiter
60	4038	Crv	ScG	12 01.9	−18 52	11.3	2.6×1.8	The antennae
61	4039	Crv	ScG	12 01.9	−18 53	13.0	3.2×2.2	The antennae
62	247	Cet	SG	00 47.1	−20 46	8.9	20×7	
63	7293	Aqr	PN	22 29.6	−20 48	6.5	13	Helix nebula
64	2362	CMa	OC	07 18.8	−24 57	4.1	8	Tau CMa cluster
65	253	Scl	SG	00 47.6	−25 17	7.1	25×7	Sculptor galaxy
66	5694	Hya	GC	14 39.6	−26 32	10.2	3.6	
67	1097	For	SBbG	02 46.3	−30 17	9.2	9×6	
68	6729	CrA	BN	19 01.9	−36 57	9.7	1.0	R CrA nebula
69	6302	Sco	PN	17 13.7	−37 06	12.8	0.8	Bug nebula
70	300	Scl	SdG	00 54.9	−37 41	8.1	20×13	
71	2477	Pup	OC	07 52.3	−38 33	5.8	27	
72	55	Scl	SBG	00 14.9	−39 11	8.2	32×6	Brightest in Scl cluster
73	1851	Col	GC	05 14.1	−40 03	7.3	11	
74	3132	Vel	PN	10 07.7	−40 26	8.2	0.8	
75	6124	Sco	OC	16 25.6	−40 40	5.8	29	
76	6231	Sco	OC	16 54.0	−41 48	2.6	15	
77	5128	Cen	Peculiar galaxy	13 25.5	−43 01	7.0	18×14	Cen A radio source
78	6541	CrA	GC	18 08.0	−43 42	6.6	13	
79	3201	Vel	GC	10 17.6	−46 25	6.7	18	
80	5139	Cen	GC	13 26.8	−47 29	3.6	36	Omega centauri
81	6352	Ara	GC	17 25.5	−48 25	8.1	7	

(continued)

(continued)

C#	NGC/IC	Con.	Type	R. A. h m	Dec. ° ′	Mag.	Size ()	Description
82	6193	Ara	OC	16 41.3	−48 46	5.2	15	
83	4945	Cen	SBcG	13 05.4	−49 28	9.5	20×4	
84	5286	Cen	GC	13 46.4	−51 22	7.6	9	
85	IC 2391	Vel	OC	08 40.2	−53 04	2.5	50	o (Omicron) Vel cluster
86	6397	Ara	GC	17 40.7	−53 40	5.6	26	
87	1261	Hor	GC	03 12.3	−55 13	8.4	7	
88	5823	Cir	OC	15 05.7	−55 36	7.9	10	
89	6087	Nor	OC	16 18.9	−57 54	5.4	12	S Nor cluster
90	2867	Car	PN	09 21.4	−58 19	9.7	0.2	
91	3532	Car	OC	11 06.4	−58 40	3.0	55	
92	3372	Car	BN	10 43.8	−59 52	6.2	120×120	Eta carinae nebula
93	6752	Pav	GC	19 10.9	−59 59	5.4	20	
94	4755	Cru	OC	12 53.6	−60 20	4.2	10	Jewel box cluster
95	6025	TrA	OC	16 03.7	−60 30	5.1	12	
96	2516	Car	OC	07 58.3	−60 52	3.8	30	
97	3766	Cen	OC	11 36.1	−61 37	5.3	12	
98	4609	Cru	OC	12 42.3	−62 58	6.9	5	
	–	Cru	DN	12 53.0	−63 00	–	400×300	Coal sack
100	IC 2944	Cen	OC	11 36.6	−63 02	4.5	15	−(Lambda) Cen cluster
101	6744	Pav	SBbG	19 09.8	−63 51	9.0	16×10	
102	IC 2602	Car	OC	10 43.2	−64 24	1.9	50	÷(Theta) Car cluster
103	2070	Dor	BN	05 38.7	−69 06	1.0	40×25	Tarantula Neb. in LMC
104	362	Tuc	GC	01 03.2	−70 51	6.6	13	
105	4833	Mus	GC	12 59.6	−70 53	7.3	14	
106	104	Tuc	GC	00 24.1	−72 05	4.0	31	47 tucanae
107	6101	Aps	GC	16 25.8	−72 12	9.3	11	
108	4372	Mus	GC	12 25.8	−72 40	7.8	19	
109	3195	Cha	PN	10 09.5	−80 52	–	0.6	

Key to Object Types: *BN* Bright nebula, *GC* Globular cluster, *OC* Open cluster, *EG* Elliptical (type) galaxy, *DN* Dark nebula, *IG* Irregular galaxy, *PN* Planetary nebula, *SN* Supernova remnant, *SG* Spiral (type) galaxy

Appendix F

Selected Non-Messier Catalog NGC Objects

The Nexstar+ hand control software contains J.L.E. Dreyer's New General Catalogue of Nebulas and Cluster of Stars (NGC) and the two supplements, the Index Catalogues (IC) in its database.

The NGC list, compiled in 1888, contains 7840 objects. The two supplements were published in 1895 and 1908 respectively, with the first containing 1520 additional deep sky objects and the second containing 3866 additional IC objects.

Many of the objects listed in the NGC/IC database are difficult or not visible visually through the size of telescope represented by the NexStar Evolution family of telescopes. It is possible to detect and image the fainter objects through long exposures or multiple exposures and stacking techniques.

Unlike the much shorter Messier Catalog of Appendix D and the Caldwell Catalog of Appendix E, it is impractical to list the thousands of NGC/IC objects in this book.

However, here are some NGC and IC objects that are recommended. This list is the called the SAA 100.

A discussion thread began in 2000 in the sci.astro.amateur (SAA) newsgroup when a question was posted: "What are your favorite non-Messier objects for 8–12″ telescopes?" The SAA Newsgroup participants responded enthusiastically to the question, posting many messages nominating a wide variety of objects.

These objects are easily accessed using the NexStar+ hand control. Use the NGC or IC nomenclature for GoTo searches.

If the SkyPortal app is used, enter the RA and Dec coordinates for accurate GoTo searches.

© Springer International Publishing Switzerland 2016   183
J.L. Chen, *The NexStar Evolution and SkyPortal User's Guide*,
The Patrick Moore Practical Astronomy Series,
DOI 10.1007/978-3-319-32539-2

The table below lists the SAA 100 in rank order by number of votes received. About half the objects received only one vote each; these are listed alphabetically at the end of the list.

F.1 SAA 100

Object	Type	Con	VisualMag	Size	RA	Dec	Pop. name	Notes
NGC 253	Gal	Scl	7.2	25.0′ × 7.0′	00 h 47 min 35 s	−25° 17′ 01″		Edge-on spiral
NGC 4565	Gal	Com	9.6	15.5′ × 1.9′	12 h 36 min 20 s	+25° 59′ 23″	Bernice's hair clip	Classic edge-on spiral with dust lane
NGC 6960	SNR	Cyg		70.0′ × 6.0′	20 h 45 min 38 s	+30° 43′ 20″	Western veil	
NGC 6992	SNR	Cyg		25.0′ × 20.0′	20 h 56 min 14 s	+31° 04′ 20″	Eastern veil	
NGC 869	OC	Per	5.3	30.0′	02 h 19 min 03 s	+57° 08′ 58″	Double cluster	w/NGC 884
NGC 884	OC	Per	6.1	30.0′	02 h 22 min 27 s	+57° 06′ 57″	Double cluster	w/NGC 869
NGC 457	OC	Cas	6.4	13.0′	01 h 19 min 10 s	+58° 20′ 02″	owl cluster	
NGC 5139	GC	Cen	3.7	36.3′	13 h 26 min 46 s	−47° 28′ 45″	Omega centauri	Best GC in the sky
NGC 7293	PN	Aqr	6.3	16.0′ × 12.0′	22 h 29 min 40 s	−20° 47′ 23″	Helical nebula	
NGC 7789	OC	Cas	6.7	16.0′	23 h 57 min 04 s	+56° 44′ 09″		
NGC 2237	BN	Mon	5.5	70.0′ × 80.0′	06 h 32 min 19 s	+04° 59′ 03″	Rosette nebula	OC NGC 2244 embedded in nebula
NGC 2244	OC	Mon	4.8	24.0′	06 h 32 min 25 s	+04° 52′ 03″		Involved with Rosette Neb. (NGC 2237)
NGC 2359	BN	CMa		8.0′	07 h 17 min 48 s	−13° 12′ 54″	Thor's helmet; duck nebula	Wolf-Rayet remnant
NGC 2392	PN	Gem	8.6	47.0″ × 43.0″	07 h 29 min 10s	+20° 54′ 42″	Eskimo nebula; clown face	
NGC 3242	PN	Hya	8.6	40.0″ × 35.0″	10 h 24 min 48 s	−18° 38′ 14″	Ghost of jupiter	
NGC 6543	PN	Dra	8.3	22.0″ × 16.0″	17 h 58 min 36 s	+66° 38′ 17″	Cat's Eye nebula	
NGC 4631	Gal	CVn	9.2	17.0′ × 3.5′	12 h 42 min 11 s	+32° 32′ 42″		Same LP field as NGC 4656
NGC 4656	Gal	CVn	10.5	22.0′ × 3.0′	12 h 43 min 58 s	+32° 10′ 21″		Same LP field as NGC 4631
NGC 5128	Gal	Cen	6.8	18.2′ × 14.5′	13 h 25 min 29 s	−43° 01′ 07″	Centaurus A	Strong radio source
NGC 6781	PN	Aql	11.8	1.9′ × 1.8′	19 h 18 min 28 s	+06° 32′ 46″		
NGC 6826	PN	Cyg	8.8	27.0″ × 24.0″	19 h 44 min 53 s	+50° 31′ 42″	Blinking planetary	
NGC 7009	PN	Aqr	8.3	28.0″ × 23.0″	21 h 04 min 15 s	−11° 21′ 49″	Saturn nebula	
Abell 1656	Gal cluster	Com	11.0	120.0′	12 h 59 min 48 s	+27° 59′ 04″	Coma gal cluster	Tough in 8–12″ aperture
NGC 1023	Gal	Per	9.4	9.0′ × 4.0′	02 h 40 min 27 s	+39° 03′ 47″		
NGC 2362	OC	CMa	4.1	8.0′	07 h 18 min 48 s	−24° 56′ 51″		
NGC 2403	Gal	Cam	8.5	17.8′	07 h 36 min 55 s	+65° 35′ 42″		

(continued)

(continued)

Object	Type	Con	VisualMag	Size	RA	Dec	Pop. name	Notes
NGC 4038	Gal	Cor	10.3	2.6′ × 1.8′	12 h 01 min 53 s	−18° 51′ 55″	The antennae; ringtail gal	Interacting with NGC 4039
NGC 4039	Gal	Cor	10.6	3.2′ × 2.2′	12 h 01 min 54 s	−18° 53′ 07″	The antennae	Interacting with NGC 4038
NGC 5907	Gal	Dra	10.3	12.8′ × 1.8′	15 h 15 min 52 s	+56° 19′ 48″		
NGC 6369	PN	Oph	11.0	30.0″ × 29.0″	17 h 29 min 22 s	−23° 45′ 37″		
NGC 663	OC	Cas	7.1	16.0′	01 h 46 min 04 s	+61° 15′ 00″		
NGC 654	OC	Cas	6.5	5.0′	01 h 44 min 10s	+61° 53′ 00″		
NGC 659	OC	Cas	7.9	5.0′	01 h 44 min 16 s	+60° 42′ 00″		
NGC 7000	BN	Cyg		175.0′ × 110.0′	20 h 58 min 32 s	+44° 33′ 21″	North American nebula	Large; often easier in binoculars than telescope
NGC 7331	Gal	Peg	9.5	11.4′ × 4.0′	22 h 37 min 08 s	+34° 25′ 27″	Little and gal	
NGC 7662	PN	And	8.6	17.0″ × 14.0″	23 h 25 min 57 s	+42° 32′ 44″	Blue snowball nebula	
B 59, 65-7	DN	Oph		300.0′	17 h 21 min 02 s	−26° 59′ 58″	Pipe nebula (stem)	
B 78	DN	Oph		200.0′	17 h 33 min 02 s	−25° 59′ 58″	Pipe nebula (bowl)	
IC 1396	BN	Cep	3.5	154.0′ × 140.0′	21 h 39 min 09 s	+57° 46′ 58″		
IC 418	PN	Lep	10.7	14.0″ × 11.0″	05 h 27 min 30 s	−12° 41′ 32″		
IC 4665	OC	Oph	4.2	41.0′	17 h 46 min 20 s	+05° 43′ 08″		
Mel 111	OC	Com	1.8	275.0′	12 h 25 min 00 s	+26° 00′ 07″	Coma berenices star cluster	
Mel 20	OC	Per	1.2	185.0′	03 h 22 min 03 s	+48° 59′ 56″	Alpha Persei Association	
NGC 1502	OC	Cam	6.9	8.0′	04 h 07 min 45 s	+62° 19′ 49″		Near SE end of Kemble's Cascade
NGC 1528	OC	Per	6.4	24.0′	04 h 15 min 24 s	+51° 13′ 49″		
NGC 1907	OC	Aur	8.2	7.0′	05 h 28 min 00 s	+35° 18′ 53″		
NGC 1973	BN	Ori		5.0′ × 5.0′	05 h 35 min 09 s	−04° 43′ 56″	Part of running man nebula	
NGC 1975	BN	Ori		10.0′ × 5.0′	05 h 35 min 21 s	−04° 40′ 56″	Part of running man nebula	
NGC 1977	BN	Ori		20.0′ × 10.0′	05 h 35 min 27 s	−04° 49′ 56″	Part of running man nebula	42 orionis nebula
NGC 2070	BN	Dor	8.3	5.0′	05 h 38 min 39 s	−69° 04′ 51″	Tarantula nebula	In Lg. magellanic cloud
3C 273	Quasar	Vir	12.0		12 h 29 min 06 s	+02° 03′ 01″		Brightest quasar; most remote object visible in modest amateur telescopes (approximately two billion light years)

Object	Type	Const.	Mag	Size	RA	Dec	Common name	Notes
Albireo	Star	Cyg	3.1		19 h 30 min 45 s	+27° 57' 55"		Superb double star; blue–white/yellow
Cr 399	Asterism	Vul	3.6	60.0'	19 h 25 min 26 s	+20° 11' 18"	Brocchi's cluster; the coathanger	Once assumed to be a OC; data from Hipparcos spacecraft shows it to be a chance alignment of stars
Fornax Gal. cluster	Gal cluster	For		3°×2°	03 h 38 min 31 s	−35° 26' 40"		Approximately 20 galaxies brighter than mag. 13
Kemble's cascade	Asterism	Cam			03 h 57 min 30s	+63° 04' 13"		First described by Canadian amateur Lucian J. Kemble; beautiful chain of about 20 mag. 5…9 stars; coordinates are for SAO 12969, a mag. 5 star in the middle of the Cascade
King 10	OC	Cep		3.0'	22 h 54 min 58 s	+59° 10' 16"		
Markarian's chain	Gal chain	Vir			12 h 25 min 04 s	+12° 53' 16"		String of bright galaxies; covers 3° of sky, starting with M84 & M86 in Virgo, ending with NGCs 4459 & 4474 in Coma Berenices; coordinates are for M 84
Mel 25	OC	Tau	0.5	330.0'	04 h 27 min 02 s	+16° 00' 03"	Hyades	Aldebaran not a member
NGC 104	GC	Tuc	4.0	30.9'	00 h 24 min 10s	−72° 04' 37"	47 tucanae	
NGC 1535	PN	Eri	10.4	20.0"×17.0"	04 h 14 min 16 s	−12° 44' 16"		Multiple shells
NGC 2158	OC	Gem	8.6	5.0'	06 h 07 min 33 s	+24° 05' 56"		
NGC 2169	OC	Ori	5.9	7.0'	06 h 08 min 27 s	+13° 56' 59"	"37" cluster	
NGC 2174	BN	Ori		25.0'×20.0'	06 h 10 min 01 s	+20° 33' 58"		
NGC 2232	OC	Mon	3.9	30.0'	06 h 26 min 37 s	−04° 44' 54"		
NGC 225	OC	Cas	7.0	12.0'	00 h 43 min 28 s	+61° 47' 06"		
NGC 2261	BN	Mon		2.0'×1.0'	06 h 39 min 13 s	+08° 44' 01"	Hubble's variable nebula	
NGC 2264	OC	Mon	3.9	30.0'×60.0'	06 h 40 min 58 s	+09° 53' 42"	Christmas tree cluster; cone nebula	Includes naked-eye S Mon (15 Mon)
NGC 2301	OC	Mon	6.0	12.0'	06 h 51 min 49 s	+00° 28' 04"		

(continued)

(continued)

Object	Type	Con	VisualMag	Size	RA	Dec	Pop. name	Notes
NGC 2360	OC	CMa	7.2	13.0′	07 h 17 min 48 s	−15° 36′ 53″		
NGC 2438	PN	Pup	11.0	1.1′	07 h 41 min 51 s	−14° 44′ 06″		In foreground of M 46
NGC 2467	BN	Pup	7.1	15.0′	07 h 52 min 30 s	−26° 22′ 52″		Use UHC or O-III filter; includes loose cluster of mag. 8–12 stars
NGC 247	Gal	Cet	9.1	20.0′×7.0′	00 h 47 min 11 s	−20° 45′ 21″		
NGC 2841	Gal	Uma	9.2	7.4′×3.5′	09 h 22 min 01 s	+50° 58′ 21″		
NGC 2903	Gal	Leo	9.0	13.3′×6.0′	09 h 32 min 10 s	+21° 29′ 58″		
NGC 3115	Gal	Sex	8.9	8.3′×3.2′	10 h 05 min 14 s	−07° 43′ 06″	Spindle gal	
NGC 3372	BN	Car		120.0′×120.0′	10 h 43 min 47 s	−59° 52′ 01″	Eta carina nebula	
NGC 3532	OC	Car	3.0	55.0′	11 h 06 min 23 s	−58° 40′ 03″		
NGC 3766	OC	Cen	5.3	12.0′	11 h 36 min 05 s	−61° 37′ 04″		
NGC 3877	Gal	Uma	11.0	5.6′×1.2′	11 h 46 min 07 s	+47° 29′ 37″		
NGC 40	PN	Cep	10.7	1.0′×0.7′	00 h 13 min 08 s	+72° 31′ 47″		
NGC 4244	Gal	CVn	10.4	18.5′×2.3′	12 h 17 min 29 s	+37° 48′ 28″		
NGC 4361	PN	Cor	10.3	1.3′	12 h 24 min 30 s	−18° 47′ 38″		
NGC 4526	Gal	Vir	9.7	7.0′×2.7′	12 h 34 min 03 s	+07° 42′ 03″	Lost gal	
NGC 4567	Gal	Vir	11.3	3.0′×2.5′	12 h 36 min 33 s	+11° 15′ 33″	Siamese twins	Overlaps NGC 4568
NGC 4568	Gal	Vir	10.8	5.1′×2.4′	12 h 36 min 35 s	+11° 14′ 17″	Siamese twins	Overlaps NGC 4567
NGC 4755	OC	Cru	4.2	10.0′	12 h 53 min 35 s	−60° 20′ 08″	Jewel box cluster; kappa crucis	
NGC 5746	Gal	Vir	10.3	7.4′×1.1′	14 h 44 min 57 s	+01° 57′ 20″		
NGC 6210	PN	Her	9.7	20.0′×13.0″	16 h 44 min 30 s	+23° 48′ 46″		
NGC 6231	OC	Sco	2.6	15.0′	16 h 54 min 01 s	−41° 48′ 06″	Table of scorpius	Zeta Sco complex
NGC 6397	GC	Ara	5.7	25.7′	17 h 40 min 43 s	−53° 40′ 33″		One of the nearest Globulars
NGC 6545	Gal	Pav	13.2	1.0′×0.9′	18 h 12 min 18 s	−63° 46′ 45″	Needle galaxy	
NGC 6572	PN	Oph	9.0	15.0′×12.0″	18 h 12 min 09 s	+06° 51′ 01″		
NGC 6633	OC	Oph	4.6	27.0′	18 h 27 min 43 s	+06° 34′ 14″		

Best Non-Messier Objects, in Rank Order

Name	Type	Const	Mag	Size	RA	Dec	Notes
NGC 6819	OC	Cyg	7.3	5.0'	19 h 41 min 20 s	+40° 11' 22"	
NGC 6885	OC	Vul	8.1	7.0'	20 h 12 min 02 s	+26° 29' 20"	
NGC 6888	BN	Cyg		20.0'×10.0'	20 h 12 min 14 s	+38° 20' 21"	Crescent nebula
NGC 6939	OC	Cep	7.8	8.0'	20 h 31 min 27 s	+60° 38' 22"	
NGC 752	OC	And	5.7	50.0'	01 h 57 min 51 s	+37° 41' 05"	
NGC 891	Gal	And	9.9	14.0'×3.0'	02 h 22 min 36 s	+42° 20' 50"	Edge-on spiral w/prominent dust lane
Stock 2	OC	Cas	4.4	60.0'	02 h 15 min 04 s	+59° 15' 58"	Muscleman cluster

Best Non-Messier Objects, in Rank Order Object types: *Gal* galaxy, *OC* open cluster, *GC* globular cluster, *PN* planetary nebula, *BN* bright nebula, *DN* dark nebula, *SNR* supernova remnant

Appendix G

The Herschel 400

Listed here are the Astronomical League's listing of 400 Herschel objects, for which the AL grants the *Herschel Award*. This list was selected and compiled by Brenda F. Guzman (Branchett), Lydel Guzman, Paul Jones, James Morrison, Peggy Taylor and Sara Saey of the Ancient City Astronomy Club in St. Augustine. It is a subset of William Herschel's work of 2514 deep sky objects. Many of these objects are included in the Messier, Caldwell, and SAA100 catalog.

Many of the objects listed in the Herschel 400 are difficult or not visible visually through the size of telescope represented by the NexStar Evolution family of telescopes. It is possible to detect and image the fainter objects through long exposures or multiple exposures and stacking techniques.

These objects are easily accessed using the NexStar+ hand control. Use the NGC or IC nomenclature for GoTo searches.

If the SkyPortal app is used, enter the RA and Dec coordinates for accurate GoTo searches.

© Springer International Publishing Switzerland 2016
J.L. Chen, *The NexStar Evolution and SkyPortal User's Guide*,
The Patrick Moore Practical Astronomy Series,
DOI 10.1007/978-3-319-32539-2

X	NGC #	H #	Const.	Mag.	Type	Size	Season	RA+Dec	Comment
*	205	18-5	And	8	G	17.4	F	0040.4+4141	M110
	404	224-2	And	10.7	G	2.1×2.0	F	0109.4+3543	el circular beautiful brt nucleu
*	752	32-7	And	5.7	OC	50	F	0157.8+3741	70 stars large scattered outward
*	7662	18-4	And	9	PN	32×23	F	2325.9+4233	Blue–green oval bright dense
	7686	69-8	And	5.6	OC	7.4	F	2330.2+4908	35 stars loose poor 1 red/orange
*	7009	1-4	Aqr	8.4	PN	44×2	F	2104.2−1122	Saturn Nebula
	7606	104-1	Aqr	10.8	G	4.4×1.5	F	2319.1−0846	sp elongated elusive large
	7723	110-1	Aqr	11.1	G	2.2×1.6	F	2338.9−1258	Bsp faint small elusive
	7727	111-1	Aqr	10.7	G	2.7	F	2339.9−1218	sp elusive slightly elongated
	772	112-1	Ari	10.9	G	5×3	F	0159.3+1901	sp elongated elusive
	129	79-8	Cas	6.5	OC	21	F	0029.9+6014	50 stars bright rich 6 mag star
	136	35-6	Cas	11.3	OC	1	F	0031.5+6132	10 stars rich resolvable elongat
	185	707-2	Cas	9.2	G	11.5	F	0039.0+4820	el circular real challenge
	225	78-8	Cas	12	OC	7.5	F	0043.4+6147	21 stars large rich
	278	159-1	Cas	10.9	G	1.3×1.3	F	0052.1+4733	el round bright nucleus
*	381	64-8	Cas	9.2	OC	6	F	0108.3+6135	24 stars yellow+red–orange star
	436	45-7	Cas	8.8	OC	4	F	0115.6+5849	25 stars loose poor large brt
*	457	42-1	Cas	6.4	OC	10	F	0119.1+5820	50 stars brt rich scattered bug?
*	559	48-7	Cas	9.5	OC	4	F	0129.5+6318	60 righ bright loosely scattered
*	637	49-7	Cas	8.2	OC	3	F	0142.9+6400	20 stars compact chain of stars
*	654	46-7	Cas	6.5	OC	5	F	0144.1+6153	50 stars loose poor a brt star
*	659	65-8	Cas	7.9	OC	5	F	0144.2+6042	20 stars loose poor near 663
*	663	31-6	Cas	7.1	OC	16	F	0146.0+6115	80 stars rich resolvable nebulous
*	7789	30-6	Cas	6.7	OC	16	F	2357.0+5644	Large rich circular sl. nebulous
*	7790	56-7	Cas	8.5	OC	17	F	2358.4+6113	25 stars scattered cir nebulous
	40	58-4	Cep	10.2	PN	0.6	F	0013.0+7232	Easy round bright greenish
	7142	66-7	Cep	9.3	OC	11	F	2145.9+6548	35 stars rectangular
*	7160	67-8	Cep	6.6	OC	7	F	2153.7+6236	25 bright scattered rich resolved.
*	7380	77-8	Cep	7.2	OC	10	F	2247.0+5806	50 stars loose poor

	NGC		Con	Mag	Type	Size		Coordinates	Description
	7510	44-7	Cep	7.9	OC	2	F	2311.5+6034	20 stars loose poor arrowhead
	157	3-2	Cet	10.4	G	2.8×2.1	F	0034.8−0824	sp oval 2 stars in it brt nucleus
*	246	25-5	Cet	8.5	PN	3.8	F	0047.0−1153	Large oval difficult 2 field star
	247	20-5	Cet	8.9	G	20	F	0047.1−2046	sp elusive elongated
	584	100-1	Cet	10.8	G	3.8	F	0131.3−0652	el small dim round fair nucleus
	596	4-2	Cet	10.9	G	3.5	F	0132.9−0702	el very small circular
	615	282-2	Cet	11.6	G	4	F	0135.1−0720	sp small elusive
	720	105-1	Cet	10.5	G	4.4	F	0153.0−1344	el small round brt nucleus
	779	101-1	Cet	11.3	G	3.7×.9	F	0159.7−0558	Bsp edge on extremely elongated
*	7000	37-5	Cyg		DN	120×100	F	2058.8+4420	North American nebula
	7008	192-1	Cyg	13.5	PN	86×96	F	2100.6+5433	Seems brighter than 13.5
	7044	24-6	Cyg	11.3	OC	3.5	F	2112.9+4129	40 scattered stars
	7062	51-7	Cyg	8.3	OC	5	F	2123.2+4623	30 stars inter. rich
	7086	32-6	Cyg	8.4	OC	8	F	2130.5+5135	50 stars large scattered
	7128	40-7	Cyg	9.7	OC	2	F	2144.0+5329	20 stars loose poor
	7006	52-1	Del	10.3	GC	1.1	F	2101.5+1611	Faint
*	7209	53-7	Lac	6.7	OC	20	F	2205.2+4630	50 stars bright rich resolvable
*	7243	75-8	Lac	6.4	OC	20	F	2215.3+4953	50 stars loose scattered
*	7296	41-7	Lac	9.4	OC	4	F	2228.2+5217	15 stars compact elusive
*	7217	207-2	Peg	10.2	G	2.6×2.3	F	2207.9+3122	sp round
	7331	53-1	Peg	9.7	G	10×2	F	2237.1+3425	sp faint edge on
	7448	251-2	Peg	11.2	G	2×1	F	2300.1+1559	sp elongated averted vision
	7479	55-1	Peg	11.6	G	3.4×2.6	F	2304.9+1219	Bsp slightly elongated elusive
	651	193-1	Per	11	PN	157×87	F	0142.3+5135	M76
*	488	252-3	Psc	10.3	G	4.2×3.3	F	0121.8+0515	sp circular brt nucleus
*	524	151-1	Psc	10.6	G	3.2	F	0124.8+0932	el small round brt nucleus
*	253	1-5	Scl	7.1	G	25.1	F	0047.6−2517	sp elongated brt center
*	288	20-6	Scl	8.1	GC	13.8	F	0052.8−2635	Large bright nucleus
	613	281-1	Scl	10.2	G	4×2	F	0134.3−2925	Bsp slightly elongated

(continued)

(continued)

X	NGC #	H #	Const.	Mag.	Type	Size	Season	RA+Dec	Comment
*	598	17-5	Tri	5.7	G	62	F	0133.9+3039	M33
	6755	19-7	Aql	9	OC	10	S	1907.8+0414	35 stars large loose fairly rich
	6756	62-7	Aql	10.7	OC	3	S	1908.7+0441	Very difficult like a globular
	6781	743-3	Aql	11	PN	106	S	1918.4+0633	Circular nebulous no color
*	5466	9-6	Boo	8.5	GC	5	S	1405.5+2832	Large diffuse faint oval elusive
	5557	99-1	Boo	11.6	G	.9×.8	S	1418.4+3630	el small brt nucleus nebulosity
	5676	189-1	Boo	11.2	G	3×1	S	1432.8+4948	sp very faint large oval
	5689	188-1	Boo	11.4	G	2×.5	S	1435.5+4845	sp faint slight elongated elus
*	6939	42-6	Cep	10	OC	5	S	2031.4+6038	Very rich small appears nebulous
*	6946	76-4	Cep	10.5	G	9×7.5	S	2034.8+6009	sp oval large faint wispy disk
*	6826	73-4	Cyg	8.8	PN	22×24	S	1944.8+5031	Small fuzzy round some blue
	6834	16-8	Cyg	10.3	OC	4	S	1952.2+2925	15 stars a string scattered star
	6866	59-7	Cyg	9	OC	6	S	2003.7+4400	36 stars rich brt stars large
	6910	56-8	Cyg	7.5	OC	8	S	2023.1+4047	40 stars rich fairly brt large
	6905	16-4	Del	12	PN	44×37	S	2022.4+2007	Easy oval silver disk
*	6934	103-1	Del	10	GC	1.5	S	2034.2+0723	Circular fairly brt nucleus
*	5866	215-1	Dra	10.8	G	2.8×1	S	1506.5+5546	el M102
	5907	759-2	Dra	11.3	G	11.1×.7	S	1515.9+5619	sp needle very long central bulge
	5982	764-2	Dra	10.9	G	1.2×.8	S	1538.7+5921	el round fuzzy very small
	6543	37-4	Dra	8.8	PN	22	S	1758.6+6638	Oval brt opaque blue–green
*	6207	701-2	Her	11.3	G	2×1.1	S	1643.1+3650	sp oval elusive near M13 difficult
	6229	50-4	Her	8.7	GC	1.2	S	1647.0+4732	Faint oval brt central area
	5694	196-2	Hyd	11	GC	2.2	S	1439.6−2632	Small fuzzy smudge brt nucleus
*	5897	19-6	Lib	11	GC	7.3	S	1517.4−2101	Oval partly resolved elusive
*	6171	40-6	Oph	9.2	GC	2.2	S	1632.5−1303	M107
	6235	584-2	Oph	10.4	GC	1.9	S	1653.4−2211	Faint no distinct nucleus diff.
*	6284	11-6	Oph	10.5	GC	1.5	S	1704.5−2446	Easy round fairly bright small

	6287	195-2	Oph	9.9	GC	1.7	S	1705.2 − 2242	Circular brt nucleus no resolve
*	6293	12-6	Oph	9.5	GC	1.9	S	1710.2 − 2634	Easy circular small brt nucleus
*	6304	147-1	Oph	9.8	GC	1.6	S	1745.5 − 2928	Circular brt nucleus no resolve
	6316	45-1	Oph	10.0	GC	1.1	S	1716.6 − 2808	Oval small bright
*	6342	149-1	Oph	10.0	GC	0.5	S	1721.2 − 1935	Small faint central area brt
	6355	46-1	Oph	10.5	GC	1	S	1724.0 − 2621	Circular elus extremely faint
*	6356	48-1	Oph	9.5	GC	1.7	S	1723.6 − 1749	Fairly large brt nucleus no resolved
*	6369	11-4	Oph	9.9	PN	28	S	1729.3 − 2346	Round blue–green Pipe Nebula
	6401	44-1	Oph	11	GC	1	S	1738.6 − 2355	brt circular no resolution
	6426	587-2	Oph	11.5	GC	1.3	S	1744.9 + 0300	Faint elus resembles a nebula
	6517	199-2	Oph	10.5	GC	0.4	S	1801.8 − 0858	Faint elus brt central area
*	6633	72-8	Oph	5.5	OC	20	S	1827.7 + 0634	44 stars rich resolved. a red star
	6144	10-6	Sco	10.5	GC	3.3	S	1627.3 − 2602	Faint oval fairly large elusive
*	6664	12-8	Sct	8.9	OC	18	S	1836.7 − 0813	25 stars poor loose scattered
*	6712	47-1	Sct	10	GC	2.1	S	1853.1 − 0842	Round opaque touches PN IC1298
	6118	402-2	Ser	11.5	G	4.3 × 1.3	S	1621.8 − 0217	sp Blinking Galaxy elus near mag 6
	6440	150-1	Sgr	10.4	GC	0.7	S	1748.9 − 2022	brt nucleus no resolution
	6445	586-2	Sgr	11	PN	38 × 29	S	1749.2 − 2001	Circular brt blue–green
	6451	13-6	Sgr	8.5	OC	6	S	1750.7 − 3013	Tightly grouped large faint in
*	6514	41-1	Sgr	6.9	OC	29 × 27	S	1802.3 − 2302	M20 Trifid Nebula star cluster
	6520	7-7	Sgr	8.1	OC	5	S	1803.4 − 2754	25 stars easily resolvable
	6522	49-1	Sgr	9.5	GC	0.7	S	1803.6 − 3002	Faint small some resolution
	6528	200-2	Sgr	10.5	GC	0.5	S	1804.8 − 3003	Near 6522 some resolution
	6540	198-2	Sgr	11	OC	0.5	S	1806.3 − 2749	20 stars rich small diffuse
	6544	197-2	Sgr	10	GC	1	S	1807.3 − 2500	brt oval resolution good edges
	6553	12-4	Sgr	10	GC	1.7	S	1809.3 − 2554	Righ tight fuzzy resolution
	6568	30-7	Sgr	8.5	OC		S	1812.8 − 2136	33 stars loose poor wide
	6569	201-2	Sgr	10.4	GC	1.4	S	1813.6 − 3150	Easy faint brt central area
	6583	31-7	Sgr	11.5	OC	1.5	S	1815.8 − 2208	Elus faint rich like a globular

(continued)

(continued)

X	NGC #	H #	Const.	Mag.	Type	Size	Season	RA+Dec	Comment
	6624	50-1	Sgr	9.5	GC	2	S	1823.7−3022	Fairly large brt nucleus
	6629	204-2	Sgr	10.6	PN	16×14	S	1825.7−2312	Circular no color brt nucleus
*	6638	51-1	Sgr	10.2	GC	1.4	S	1830.9−2530	Small faint elusive
	6642	205-2	Sgr	10.5	GC	1	S	1831.9−2329	Very faint small elusive
	6645	23-6	Sgr	8.5	OC	10	S	1832.6−1645	50 stars rich compressed brt
*	6818	51-4	Sgr	10	PN	22×15	S	1944.0−1409	Diffuse fuzzy rount some blue
	5473	231-1	UMa	11.4	G	.9×.7	S	1404.7+5454	el companion to M101 round elus.
*	5474	214-1	UMa	11.4	G	4×2.9	S	1405.0+5340	sp near M101 dim diffuse elusive
	5631	236-1	UMa	11.4	G	.7×.7	S	1426.6+5635	sp small round faint difficult
	6217	280-1	UMn	11.5	G	1.8×1.2	S	1632.6+7812	sp faint sl oval elusive arms
	5566	144-1	Vir	10.4	G	5.6×1.1	S	1420.3+0356	sp edgeon faint lg brt nuc elus
	5576	146-1	Vir	11.7	G	1×.8	S	1421.1+0316	el small faint brt nuc nebulosity
*	5634	70-1	Vir	10.4	GC	1.3	S	1429.6−0559	Small fuzzy patch opaque oval
	5746	126-1	Vir	10.1	G	6.2×.8	S	1444.9+0157	sp edge-on needle faint large elus
	5846	128-1	Vir	10.5	G	.9×.9	S	1506.4+0136	el small circular fairly bright
	6802	14-6	Vul	11	OC	3.5	S	1930.6+2016	Faint small opaque disk of stars
*	6823	18-7	Vul	9.8	OC	5	S	1943.1+2318	30 stars some nebulosity
*	6830	9-7	Vul	9	OC	8	S	1951.0−2304	20 stars brt large easy
	6882	22-8	Vul	5.5	OC		S	2011.7+2633	20 stars small in a rich field
	6885	20-8	Vul	9.1	OC	20	S	2012.0+2629	42 stars large a brt star rich
*	6940	8-7	Vul	6.5	OC	10	S	2034.6+2818	100 stars very brt rich large
	5248	34-1	Boo	11.3	G	6.1×4.4	Sp	1337.5+0853	sp lg oval brt nucleus av
	2655	288-1	Cam	10.7	G	5×3.4	Sp	0855.6+7813	sp faint face-on low surf brt
	2775	2-1	Cnc	10.7	G	2.3×1.9	Sp	0910.3+0702	sp circular brt nucleus
	4147	19-1	Com	9.4	GC	1.7	Sp	1210.1+1833	Small starlike outer nebulosity
	4150	73-1	Com	11.6	G	1.2×.9	Sp	1210.6+3024	el round starlike nucl a brt sta
	4203	175-1	Com	11	G	1.8×1.5	Sp	1215.1+3312	el sl oval faint brt nucleus av

	4245	74-1	Com	11.1	G	1.5×1	Sp	1217.6+2936	sp elus starlike hint elong av
	4251	89-1	Com	10.2	G	2.3×.5	Sp	1218.1+2810	sp round starlike nebulosity
	4274	75-1	Com	10.8	G	6.7×1.3	Sp	1219.8+2937	sp lg fuzzy brt nucl elongated
	4278	90-1	Com	10.3	G	1.4×1.3	Sp	1220.1+2917	el starlike brt nucl defined
	4293	5-5	Com	11.5	G	4.6×1.6	Sp	1221.2+1823	pec edge-on granulated elus elon
	4314	76-1	Com	10.8	G	3.1×2.9	Sp	1222.6+2953	Bsp oval diffuse brter nucleus
	4350	86-2	Com	11	G	1.8×.5	Sp	1224.0+1642	el round starlike nebulosity
	4394	55-2	Com	11.2	G	2.3×2.3	Sp	1225.9+1813	Bsp round brt nuc av near M85
	4414	77-1	Com	9.7	G	3.2×1.5	Sp	1226.4+3113	sp starlike nuc extended arms
	4419	113-1	Com	11.4	G	2.2×.6	Sp	1226.9+1503	el elong brt nucl fainter arm av
	4448	91-1	Com	11.4	G	2.8×1	Sp	1228.2+2837	sp sl elong uniform av elus
*	4450	56-2	Com	10	G	3×2.5	Sp	1228.5+1705	sp round brt distinct nucleus
	4459	161-1	Com	10.9	G	1.2×1	Sp	1229.0+1359	el brt sl elong
*	4473	114-2	Com	10.1	G	1.6×.9	Sp	1229.8+1326	el round starlike nucl brt
*	4477	115-2	Com	10.7	G	2.4×2.2	Sp	1230.0+1438	sp round starlike nucl fair brt
*	4494	83-1	Com	9.6	G	1.3×1.4	Sp	1231.4+2547	el brt starlike neb round small
*	4548	120-2	Com	10.8	G	3.7×3.2	Sp	1235.4+1430	Bsp M91
*	4559	92-1	Com	10.6	G	11×4.5	Sp	1236.0+2758	sp elong uniform fairly brt
*	4565	24-5	Com	10.2	G	14.4×1.2	Sp	1236.3+2559	sp uniform needle beautiful
	4689	128-2	Com	11.5	G		Sp	1247.8+1346	sp round uniform av
*	4725	84-1	Com	8.9	G	10×5.5	Sp	1250.4+2530	sp oval starlike nucl easy
	4027	296-2	Cor	11.5	G	2.4×2	Sp	1159.5−1916	sp faint elus hint elong av
	4038	28.1-4	Cor	11.5	G	2.5×2.5	Sp	1201.9−1852	sp round brt nucleus
*	4361	65-1	Cor	10.8	PN	81	Sp	1224.5−1848	lg round fuzzy no color noted
	3962	67-1	Cra	11.3	G	1.1×.9	Sp	1154.7−1358	el round very faint brt nucleus
	4111	195-1	CVn	9.7	G	3.3×.6	Sp	1207.1+4304	el elong starlike outer arms av
	4143	54-4	CVn	11	G	1.4×.9	Sp	1209.6+4232	el round starlike nucleus small
	4151	165-1	CVn	11.6	G	2.5×1.6	Sp	1210.5+3924	pec uniform round ovalish elus
	4214	95-1	CVn	10.3	G	6.6×5.8	Sp	1215.6+3620	irr very brt large diffuse oval

(continued)

(continued)

X	NGC #	H #	Const.	Mag.	Type	Size	Season	RA+Dec	Comment
*	4258	43-55	CVn	8.6	G	19.5×7	Sp	1219.0+4718	sp M106
	4346	210-1	CVn	11.6	G	1.9×.7	Sp	1223.5+4700	el round almost starlike neb
*	4449	213-1	CVn	9.2	G	4.1×3.4	Sp	1228.2+4406	irr sl elong uniform pretty
	4485	197-1	CVn	11.6	G	1.5×.8	Sp	1230.5+4142	irr uniform brt elusive small
*	4490	198-1	CVn	9.7	G	5.6×2.1	Sp	1230.6+4138	sp lg elong uniform brt easy
*	4618	178-1	CVn	11.7	G	.5×3	Sp	1241.5+4109	sp lg sl elong uniform brt
	4631	42-4	CVn	9.3	G	12.6×1.4	Sp	1242.1+3232	sp elong uniform brt easy
	4656	176-1	CVn	11.2	G	18×2	Sp	1244.0+3210	pec lg elus needle av
	4800	211-1	CVn	11.1	G	1.2×1	Sp	1254.6+4642	sp round brt stands out well
*	5005	96-1	CVn	9.8	G	4.4×1.7	Sp	1310.9+3703	sp elong lg brt nucleus
*	5033	97-1	CVn	10.3	G	9.9×4.8	Sp	1313.4+3636	sp elong brt needle lg
*	5195	186-1	CVn	8.4	G	2×1.5	Sp	1330.0+4716	pec companion to M51 round
	5273	98-1	CVn	11.5	G	.9×.8	Sp	1342.1+3539	el brt nucleus av stands out wel
	3147	79-1	Dra	10.9	G	3×2.3	Sp	1016.9+7339	sp lg circular near 6 mag star
*	2548	22-6	Hyd	5.3	OC	30	Sp	0813.8−0548	M48
	2811	505-2	Hyd	11.7	G	1.6×.5	Sp	0916.0−1606	sp uniform brightness
*	3242	27-4	Hyd	9	PN	40×34	Sp	1024.8+1838	Nice disk blue–green fairlylarge
	3621	241-1	Hyd	10.5	G	5×2	Sp	1118.3+3249	sp lg round near some field star
	3686	160-2	Leo	11.4	G	2.4×1.8	Sp	1127.7+1713	sp strlike brt nucleus nebulosity
*	2903	56-1	Leo	9.1	G	11×4.6	Sp	0932.2−2130	sp lg elong needle easy
	2964	114-1	Leo	11	G	2.2×1.1	Sp	0942.9+3151	sp round brt nucleus faint rim
	3190	44-2	Leo	11.3	G	3×1	Sp	1018.1+2150	sp fuzzy round same field 3193
	3193	45-2	Leo	11.5	G	.9×.0	Sp	1018.4+2154	el fuzzy roundish uniform c 3190
	3226	28-22	Leo	11.5	G	1×.8	Sp	1023.4+1954	el round nebulous uniform
	3227	29-2	Leo	11.4	G	3×1.2	Sp	1023.5+1952	sp round nebulous near NGC 3226
	3377	99-2	Leo	10.5	G	1.9×1	Sp	1047.7+1359	el small brt nucleus
*	3379	17-1	Leo	9.5	G	2.2×2	Sp	1047.8+1235	el M105

*	3384	18-1	Leo	10.2	G	4.4×1.4	Sp	1048.3+1238	el round brt nucleus >NGC 3379
	3395	116-1	Leo	12	G	1.5×.9	Sp	1049.8+3259	sp elus sl elong faint nucleus
	3412	27-1	Leo	10.4	G	2.4×1.1	Sp	1050.9+1325	el circular very difficult
*	3489	101-2	Leo	11.5	G	2×.9	Sp	1100.3+1354	el round impressive brt nucleus
	3521	13-1	Leo	10.5	G	7×4	Sp	1105.8−0002	sp elongated very impressive
	3593	29-1	Leo	11.3	G	2.5×.9	Sp	1114.6+1249	sp round faint inconspicuous
*	3607	50-2	Leo	9.6	G	1.7×1.5	Sp	1116.9+1803	el like a hairy star round brt
	3608	51-2	Leo	11.1	G	1.4×1	Sp	1117.0+1809	el round small same field 3607
*	3626	52-2	Leo	10.5	G	1.6×1.1	Sp	1120.1+1821	sp starlike brt nucleus v small
*	3628	8-5	Leo	10.9	G	12×1.5	Sp	1120.3+1336	sp edge-on large beautiful
	3640	33-2	Leo	10.7	G	1.1×1	Sp	1121.1+0314	el round small faint near a star
	3810	21-1	Leo	10.8	G	3.6×2.5	Sp	1141.0+1128	sp sl lg brt nucleus
	3900	82-1	Leo	11.5	G	1.7×.8	Sp	1149.2+2701	sp starlike sl neb near 3912
	3912	342-2	Leo	11.5	G	.9×.5	Sp	1150.0+2629	sp small round uniform near 3900
	2859	137-1	LMn	10.7	G	4.4×3.5	Sp	0924.3+3431	Bsp brt nucleus fainter arm
	3245	86-1	LMn	11.2	G	1.8×.9	Sp	1027.3+2830	el round fuzzy uniform
	3277	359-2	LMn	12	G	1.1×.9	Sp	1032.9+2831	sp round brt nucleus
	3294	164-1	LMn	11.4	G	2.6×1.2	Sp	1036.3+3720	sp very difficult uniform brt
	3344	81-1	LMn	11	G	7.6×6.2	Sp	1043.5+2455	sp round faint near mag 9 star
	3414	362-2	LMn	11	G	1.4×1	Sp	1051.3+2759	Bsp starlike fuzzy nebulosity
	3432	172-1	LMn	11.4	G	5.8×.8	Sp	1052.5+3637	sp elongated uniform brt 2 stars
	3486	87-1	LMn	11	G	6.8×4.5	Sp	1000.4+2858	sp round faint fuzzy uniform
	3504	88-1	LMn	10.9	G	2.2×2.2	Sp	1100.5+2758	sp round very impressive
*	2683	200-1	Lyx	9.6	G	8×13	Sp	0852.7+3325	sp edge-on pretty needle
	2782	167-1	Lyx	11.7	G	1.8×1.6	Sp	0914.1+4007	sp starlike some nebulosity
	2527	30-8	Pup	8	OC	22	Sp	0805.3−2810	50 stars loose scattered
*	2539	11-7	Pup	8.2	OC	21	Sp	0810.7−1250	A brt center blue one 19 Pup
	2567	64-7	Pup	8.3	OC	10	Sp	0818.6−3038	50 stars brt+faint ones
	2571	39-6	Pup	7.5	OC	8	Sp	0818.9−2944	25 stars loose irregular faint

(continued)

(continued)

X	NGC #	H #	Const.	Mag.	Type	Size	Season	RA + Dec	Comment
	2613	266-2	Pyx	11	G	6.6×1.3	Sp	0833.4−2258	sp edge-on elus difficult
	2627	63-7	Pyx	8.3	OC	8	Sp	0837.3−2957	40 stars small hard distinguish
	2974	61-1	Sex	11	G	1.5×.9	Sp	0942.6−0342	sp elong fuzzy low surf brt
*	3115	163-1	Sex	9.3	G	4×1.2	Sp	1005.2+0743	el small circular
	3166	3-1	Sex	11.4	G	4.4×1.7	Sp	1013.8+0326	sp brt nucleus round field 3169
	3169	4-1	Sex	11.7	G	4×1.7	Sp	1014.2+0328	sp brt nucleus sl round
*	2681	242-1	UMa	10.4	G	2.8×2.5	Sp	0853.5+5119	sp circular faint fuzzy
	2742	249-1	UMa	11.2	G	2.5×1	Sp	0907.6+6029	sp elongated elus
	2768	250-1	UMa	10.5	G	2×1	Sp	0911.6+6002	el sl elong same field NGC 2742
	2787	216-1	UMa	10.9	G	2×1.3	Sp	0919.3+6912	sp elong near 2 brt stars
*	2841	205-1	UMa	9.3	G	6.4×2.4	Sp	0922.0+5058	sp brt elong easy
	2950	68-4	UMa	10.9	G	1.3×.9	Sp	0942.6+5851	sp starlike fuzzy nebula on edges
	2976	285-1	UMa	8.5	OC	10	Sp	0947.3+6755	50 stars rich concentrate nebulosity
	2985	78-1	UMa	10.6	G	5.5×5	Sp	0950.4+7217	sp brt nucleus easy
*	3034	79-4	UMa	8.8	G	9×4	Sp	0955.8+6941	pec M82
*	3077	286-1	UMa	10.9	G	2.3×1.9	Sp	1003.3+6844	el faint low surf brt comp. M81
	3079	47-5	UMa	11.2	G	8×1	Sp	1002.0+5541	sp edge-on faint difficult
*	3184	168-1	UMa	9.6	G	5.6×5.6	Sp	1018.3+4125	sp low surf brt difficult round
	3198	199-1	UMa	11	G	9×3.2	Sp	1019.9+4533	sp elong with 3 brt stars
*	3310	60-4	UMa	10.1	G	4×3	Sp	1035.7+5330	irr round brt nucleus
*	3556	46-5	UMa	11	G	7.7×1.3	Sp	1111.5+55.4	sp M108 edge-on
	3610	270-1	UMa	11.2	G	1.3×1	Sp	1118.4+5847	el starlike fuzzy nebulous av
	3613	271-1	UMa	11.2	G	1.6×.8	Sp	1118.6+5800	el starlike like a hairy star
	3619	244-1	UMa	11.7	G	1×1	Sp	1119.4+5646	sp starlike same field as 3613
	3631	226-1	UMa	11.2	G	4.3×3.2	Sp	1121.0+5310	sp face-on faint uniform fuzzy
	3655	5-1	UMa	11.3	G	1.2×.9	Sp	1122.9+1635	sp small fuzzy round faint
	3665	219-1	UMa	11.4	G	1.6×1.2	Sp	1124.7+3846	el round brt nucleus small

*	3675	194-1	UMa	11.5	G	4×1.7	Sp	1126.1+4335	sp faint fuzzy elongation
*	3726	730-2	UMa	10.8	G	5.7×3.4	Sp	1133.3+4702	sp round uniform largish
	3729	222-1	UMa	11.7	G	1.8×1.3	Sp	1133.8+5308	pec starlike nebulosity
	3813	94-1	UMa	11.7	G	1.7×.8	Sp	1141.3+3633	sp oval well defined 1 side brt
	3877	201-1	UMa	10.9	G	4.4×.8	Sp	1146.1+4730	sp elongated uniform brightness
*	3893	738-2	UMa	11.3	G	3.7×1.9	Sp	1148.6+4843	sp sl elong brt nucleus av
	3898	228-1	UMa	11.5	G	2.6×1	Sp	1149.2+5606	sp round brt nucleus
	3938	203-1	UMa	11.5	G	4.5×3.8	Sp	1152.8+4407	sp lg elong uniform
	3941	173-1	UMa	9.8	G	1.8×1.2	Sp	1152.9+3659	sp round nucleus nebulosity
	3945	251-1	UMa	10.8	G	5.2×2.2	Sp	1153.2+6041	Bsp diffuse strlike lg elongated
*	3949	202-1	UMa	11	G	2.3×1.1	Sp	1153.7+4752	sp oval diffuse uniform brtness
*	3953	45-5	UMa	10.7	G	5.6×2.3	Sp	1153.8+5220	sp round brt nucleus av
	3982	62-4	UMa	11.3	G	1.7×1.3	Sp	1156.5+5508	sp round elus av brt nucleus
*	3992	61-4	UMa	10.8	G	6.2×3.5	Sp	1157.6+5323	sp M109
	3998	229-1	UMa	11.3	G	3.7×1.9	Sp	1157.9+5527	el brt nucelus elong elus av
*	4026	223-1	UMa	10.7	G	3.6×.7	Sp	1156.9+5058	el round starlike nebulosity
	4036	253-1	UMa	10.7	G	2.4×.9	Sp	1201.4+6153	el oval brt nucleus small
	4041	252-1	UMa	11	G	2.4×1.8	Sp	1202.2+6208	sp round brt nucleus av
	4051	56-4	UMa	11	G	4.5×3.6	Sp	1203.2+4432	sp lg elus nebulosity av
	4085	224-1	UMa	11.8	G	2.2×.5	Sp	1205.4+5012	sp brt starlike nucleus oval
	4088	206-1	UMa	10.9	G	4.5×1.4	Sp	1205.6+5033	sp elong brt elusive large
	4102	225-1	UMa	11.8	G	2.2×1	Sp	1206.4+5243	sp hint of elong starlike brt nu
	5322	256-1	UMa	10	G	1×4×1.4	Sp	1349.3+6012	el round brt nucleus
	4030	121-1	Vir	11	G	3.1×2.2	Sp	1200.4−0106	sp round brt pretty 2 field star
	4179	9-1	Vir	11.6	G	2.7×.6	Sp	1212.9+0118	el edge-on well distinct arms
*	4216	35-1	Vir	10.4	G	7.4×.9	Sp	1215.9+1309	sp edge-on needle easy brt nucleus
*	4261	139-2	Vir	10.3	G	.9×.7	Sp	1219.4+0549	el round uniform fuzzy elus
	4273	569-2	Vir	11.6	G	1.5×1	Sp	1219.9+0521	sp starlike near group stars
	4281	573-2	Vir	11.3	G	1.1×.6	Sp	1220.4+0523	el elus starlike hint neb av

(continued)

(continued)

X	NGC #	H #	Const.	Mag.	Type	Size	Season	RA+Dec	Comment
*	4303	139-1	Vir	10.1	G	5.6×5.3	Sp	1219.4+0428	sp M61
	4365	30-1	Vir	11.1	G	1.3×1	Sp	1224.5+0719	el round faint nucleus brter
	4371	22-1	Vir	11.6	G	2.2×1.2	Sp	1224.9+1142	Bsp brt lg oval opaque brt nucleus
	4429	65-2	Vir	11.2	G	3.3×1	Sp	1227.4+1107	sp oval nucleus dominated
*	4435	28.1-1	Vir	10.3	G	1.3×.8	Sp	1227.7+1305	el elus uniform near 4438 av
*	4438	28.2-1	Vir	10.8	G	8.9×3	Sp	1227.8+1301	sp elong uniform lg av
*	4442	156-2	Vir	10.8	G	1.8×.9	Sp	1228.1+0948	el ovalish starlike neb small
*	4478	124-2	Vir	10.9	G	.8×.7	Sp	1230.3+1220	el round starlike elus neb small
	4526	31-1	Vir	10.9	G	3.3×1	Sp	1234.0+0742	el oval faint neb near 2 7mag st
	4527	37-2	Vir	11.5	G	5.3×1	Sp	1234.1+0239	sp elong uniform brightness
*	4535	500-2	Vir	11	G	6×4	Sp	1234.3+0812	sp elus elong brt nucleus
	4536	2-5	Vir	10.9	G	6.9×2.6	Sp	1234.5+0211	sp elong only can see part of it
	4546	160-1	Vir	10	G	1.8×.8	Sp	1235.5-0348	el round nebulosity uniform
	4550	36-1	Vir	11.7	G	1.4×.4	Sp	1235.5+1213	el round brt nucleus elusive
	4570	32-1	Vir	10.9	G	2.4×.5	Sp	1236.9+0715	el fuzzy round
*	4594	43-1	Vir	8.7	G	6×2.5	Sp	1240.0-1137	sp M104 The Sombrero Hat
	4596	24-1	Vir	11.4	G	1.8×2.2	Sp	1239.9+1011	Bsp round starlike nucleus
*	4636	38-2	Vir	10.4	G	1.4×1.3	Sp	1242.8+0241	el round brt nucleus
*	4643	10-1	Vir	10.6	G	1.65×.9	Sp	1243.3+0159	Bsp round starlike brt nucleus
	4654	126-2	Vir	11	G	4.2×2.2	Sp	1244.0+1308	sp lg elong stands out well
	4660	71-2	Vir	10.9	G	1.5×.8	Sp	1244.5+1111	el round starlike nebulosity
*	4665	142-1	Vir	11.1	G	3.1×2.1	Sp	1245.1+0303	Bsp almost stellar small
*	4666	15-1	Vir	11.4	G	3.8×.8	Sp	1245.1-0028	sp lg brt elong stands out well
	4697	39-1	Vir	10.5	G	2.2×1.4	Sp	1248.6-0548	el brt nucleus lg uniform
	4698	8-1	Vir	11.3	G	3×1.1	Sp	1248.4+0829	sp sl elong near 2 field stars
*	4699	129-1	Vir	9.3	G	3×2	Sp	1249.0-0840	sp lg easy brt nucleus
	4753	16-1	Vir	10.8	G	3.3×1.1	Sp	1252.4-0112	sp brt elong uniform brt

	NGC		Const	Mag	Type	Size	Code	Coord	Description
*	4754	25-1	Vir	10.5	G	1×1.2	Sp	1252.3+1119	el round brt nucl neb uniform
	4762	75-2	Vir	11	G	3.7×.4	Sp	1252.9+1114	sp elong uniform lg 3 field star
*	4781	134-1	Vir	11.2	G	2.3×1.1	Sp	1254.4−1032	sp uniform elong elus
	4845	536-2	Vir	11.5	G	4.2×.7	Sp	1258.0+0135	sp elong uniform brt easy
	4856	68-1	Vir	11.5	G	2×.7	Sp	1259.3−1502	eloval brt nucl easy to see
	4866	162-1	Vir	11.4	G	6.8×.8	Sp	1259.5+1410	sp elongated needle impress easy
	4900	143-1	Vir	11.3	G	1.7×1.5	Sp	1300.6+0230	sp starlike center oval small
	4958	130-1	Vir	10.9	G	1.7×.7	Sp	1305.8−0801	el round starlike uniform brt
	4995	42-1	Vir	11.2	G	2×1.1	Sp	1309.7−0750	sp sl elong elus uniform av
	5054	513-2	Vir	11.5	G	3.8×2.2	Sp	1317.0−1638	sp elongated cluster uniform av
	5363	6-1	Vir	10.7	G	1×1.4	Sp	1356.1+0529	wl round brt nucl stands out wel
	5364	534-2	Vir	11	G	6.2×3	Sp	1356.2+0501	sp lg elongated faint av
*	891	19-5	And	11.5	G	11.8×1.1	W0222.6+4221		sp extremely elongated edge-on
	1664	59-8	Aur	7.5	OC	15	W	0451.1+4342	25 stars faint rich elongated
	1857	33-7	Aur	8.5	OC	9	W	0520.2+3921	30 stars no shape poor loose
	1907	39-7	Aur	9.9	OC	5	W	0528.0+3519	Tight some nebulosity richish
	1931	261-1	Aur	9.5	DN	3×3	W	0531.4+3519	Starlike with nebulosity around
	2126	68-8	Aur	10	OC	6.5	W	0603.0+4954	28 stars faint poor blue white
	2281	71-8	Aur	6.9	OC	17	W	0649.3+4104	30 stars bright lg easy
*	1501	53-4	Cam	13.3, 9	PN	7	W	0407.0+6055	Round distict blue−green
*	1502	47-7	Cam	5.3	OC	7	W	0407.7+6220	15 stars compact cross-bow doubl
	1961	747-3	Cam	11.7	G	3.7×1.6	W	0542.1+6923	sp faint elus fuzzy sl elongated
	2403	44-5	Cam	8.9	G	16.8×10	W	0736.9+6536	sp lg sl elon face on faint star
	1027	66-8	Cas	7.5	OC	7	W	0242.7+6133	Large loose near 7 mag star
	908	153-1	Cet	11	G	5×2.3	W	0223.1−2114	sp elongated evenly brt large
	936	23-4	Cet	10.7	G	3.3×2.5	W	0227.6−0109	sp round starlike nucleus
	1022	102-1	Cet	11.2	G	1.8×1.4	W	0238.5−0640	Bsp small almost stellar faint

(continued)

(continued)

X	NGC #	H #	Const.	Mag.	Type	Size	Season	RA + Dec	Comment
	1052	63-1	Cet	11.2	G	1.3×1	W	0241.1−0815	el stellar small hint nebulosity
*	1055	1-1	Cet	11.5	G	6.7×1.5	W	0241.8+0026	sp edge on elus same field M77
	2204	13-7	CMj	9.1	OC	13	W	0615.7−1839	Scattered a yellow-orange star
	2354	16-7	CMj	9	OC	25	W	0714.3−2544	lg dim scattered round
*	2360	12-7	CMj	9.4	OC	12	W	0717.8−1537	Pretty brt no shape near mag 6
	2362	17-7	CMj	10.5	OC	6	W	0718.8−2457	40 stars lg brt in center temple
	1084	64-1	Eri	11	G	2.1×1.1	W	0243.0−0735	Easy round brt nucleus
*	1407	107-1	Eri	10.6	G	1.1×1.1	W	0340.2−1835	el round starlike brt nucleus
*	1535	26-4	Eri	9.3	PN	20×17	W	0414.2−1244	Small round starlike blue tint
*	2129	26-8	Gem	7.2	OC	5	W	0601.0+2318	13 stars broken chainlike
	2158	17-6	Gem	11	OC	4	W	0607.5+2406	brt lg same field as M35
*	2266	21-6	Gem	9.8	OC	5	W	0643.2+2658	30 stars 1 8mag triangular shape
	2304	2-6	Gem	10.1	OC	5.5	W	0655.0+1801	Small loose irregular
*	2355	6-6	Gem	9.5	OC	9	W	0716.9+1347	70 stars tight faint near mag 6
	2371	316-2	Gem	11	PN	54×35	W	0725.6+2929	Elongated off-white part of 2372
	2372	317-2	Gem	9.5	PN	47×43	W	0725.6+2930	Elongated off-white part ot 2371
*	2392	45-4	Gem	9.35	PN	47×43	W	0729.2+2055	Small round bluish opaque
	2395	11-8	Gem	9.4	OC	12	W	0727.1+1335	Faint elus rich scattered
*	2420	1-6	Gem	10.2	OC	7	W	0738.5+2134	20 stars some nebulosity
*	1964	21-4	Lep	11.6	G	5.4×1.1	W	0533.4−2157	sp sl elongation nucleus visible
	2419	218-1	Lyx	11.5	GC	1.7	W	0738.1+3853	Small dim stands out well
	2215	20-7	Mon	8.6	OC	8	W	0621.0−0717	10 stars round faint a brt star
	2232	25-8	Mon	4	OC	20	W	0626.6−0445	Naked eye brt stars one blue–white
*	2244	2-7	Mon	6.2	OC	24	W	0632.4+0452	16 stars within Rosette Nebula
	2251	3-8	Mon	8.5	OC	10	W	0634.7+0822	30 stars lg flat in appearance
*	2264	27-5,5-8	Mon	4.7	OC D 30	W	W	0641.1+0953	20 stars loose poor DN nebulosity

	NGC		Con	mag	type	size	W	coord	description
	2286	31-8	Mon	8	OC	15	W	0647.6−0310	50 stars loose scattered
	2301	27-6	Mon	5.8	OC	15	W	0651.8+0028	40 stars large pretty easy
*	2311	60-8	Mon	9.6	OC	7	W	0657.8−0435	25 stars brt rich easy resolvable
	2335	32-8	Mon	9.1	Oc	12	W	0706.6−1005	35 stars triangular tight group
	2343	33-8	Mon	8	OC	7	W	0708.3−1039	15 stars brt rich tight cute
*	2353	34-8	Mon	5.3	OC	20	W	0714.6−1018	25 stars loose 1 bl/wh easy
	2506	37-6	Mon	8.5	OC	10	W	0800.2−1047	50 stars rich nebulousity easy
	1788	32-5	Ori	11	DN	8×5	W	0506.9−0321	Round non uniform near 8 magstar
	1980	31-5	Ori		DN	14×14	W	0535.4−0554	Surround iota Orionis nebulosity
	1999	33-4	Ori	10	DN	16×12	W	0536.5−0642	brt nucleus hazy lg difficult
	2022	34-4	Ori	11.5	PN	28×27	W	0542.1+0905	Fuzzy oval small slight blue–gr
	2024	28-5	Ori	10.7	DN	30×30	W	0541.9−0151	Irregular patchy brt very large
*	2169	24-8	Ori	6.4	OC	5	W	0608.4+1357	15 stars brt tightly grouped
	2185	20-4	Ori	11	DN	2×2	W	0611.1−0613	brt fuzzy with NGC 2183+2184
	2186	25-7	Ori	9.5	OC	5	W	0612.2+0527	30 stars loose poor some brt
*	2194	5-6	Ori	9.2	OC	8	W	0613.8+1248	Rich slightly circular resolve
*	869	33-6	Per	4.4	OC	36	W	0219.0+5709	Double Cluster resolvable
*	884	34-6	Per	4.7	OC	36	W	0222.4+5707	Double Cluster resolvable
*	1023	156-1	Per	10.5	G	4×1.2	W	0240.3+3904	el elonated brt starlike nucl
*	1245	25-6	Per	6.9	OC	10	W	0314.7+4715	40 stars large 5 brt stars cluster
	1342	88-8	Per	7.1	OC	15	W	0331.6+3720	25 stars sl elongated poor
	1444	80-8	Per	6.4	OC	4	W	0349.4+5240	15 stars string of brt stars
*	1513	60-7	Per	8.8	OC	12	W	0410.0+4931	40 stars loose compact nucleus
*	1528	61-7	Per	6.2	OC	25	W	0415.4+5114	80 stars loosely packed rich
*	1545	85-8	Per	8	OC	18	W	0420.9+5015	25 stars loose 2 brt starswh org
	2324	38-7	Pup	8.8	OC	9	W	0704.2+0103	30 stars 5 stars=a Y loose
*	2421	67-7	Pup	9.4	OC	8	W	0736.3−2037	50 stars no shape lg fairly rich
*	2422	38-8	Pup	4.5	OC	25	W	0736.6−1430	M47
*	2423	28-7	Pup	6.9	OC	20	W	0737.1−1352	60 stars lg brt stars easy

(continued)

(continued)

X	NGC #	H #	Const.	Mag.	Type	Size	Season	RA+Dec	Comment
	2438	39-4	Pup	11.3	PN	68	W	0741.8−1444	brt obvious circular within M46
	2440	64-4	Pup	11.5	PN	54×20	W	0741.9−1813	Greenish round easy to find
	2479	58-7	Pup	9.5	OC	8	W	0755.1−1743	40 stars faint round small easy
*	2482	10-7	Pup	8.7	OC	18	W	0754.9−2418	50 stars lg rich no shape
	2489	23-7	Pup	9.4	OC	7	W	0756.2−3004	30 stars circular 2 brt st easy
	2509	1-8	Pup	9.3	OC	4	W	0800.2−1904	40 stars some brt most faint
*	1647	8-8	Tau	6	OC		W	0446.0+1904	30 stars round double stars
	1750	43-8	Tau		OC	45	W	0503.9+2339	Part of NGC 1746 3 clusters
	1817	4-7	Tau	7.9	OC	15	W	0512.2+1642	16 stars loose scattered poor

Appendix H

Current Sky Portal WiFi Module Compatible Celestron Mount Specifications

Advanced VX

Telescope type	Computerized (GOTO)
Tripod	Adjustable, stainless steel
Tripod leg diameter	2″
Mount height (max)	64″ in. (1626 mm)
Mount height (min)	44″ in. (1118 mm)
Weight capacity (max)	30 lbs
Tripod weight	18 lbs (8.16 kg)
Weight of counterweights	1 × 12 lbs
Warranty	2-year telescope warranty
Alignment procedures	AutoAlign, 2-star align, quick align, 1-star align, last alignment, solar system align
Software precision	24 bit, 0.08 calculation
Computer hand control	Double line, 16 character liquid crystal display; 19 fiber optic backlit LED buttons
Database	40,000+ objects, 100 user defined programmable objects. Enhanced information on over 200 objects
Power requirements	12 VDC 3.2 Amps
GPS	Optional SkySync GPS accessory
Alignment procedures	AutoAlign, 2-star align, quick align, 1-star align, last alignment, solar system align
Software precision	24 bit, 0.08 calculation

(continued)

© Springer International Publishing Switzerland 2016
J.L. Chen, *The NexStar Evolution and SkyPortal User's Guide*,
The Patrick Moore Practical Astronomy Series,
DOI 10.1007/978-3-319-32539-2

(continued)

Telescope type	Computerized (GOTO)
Computer hand control	Double line, 16 character liquid crystal display; 19 fiber optic backlit LED buttons
Database	40,000 + objects, 100 user defined programmable objects. Enhanced information on over 200 objects
Power requirements	12 VDC 3.2 Amps
GPS	Optional SkySync GPS Accessory

CGEM

Mount type	Computerized equatorial
Tripod	Adjustable, stainless steel
Power requirements	Car battery adapter
Mount weight	41 lbs (19 kg)
Tripod weight	17 lbs (7.71 kg)
Weight of counterweights	1 × 17 lbs
Weight (lbs)	75 lbs (34 kg)
Payload capacity	40 lbs (18 kg)
Latitude range	15–70°
Motor drive	Low cog DC servo motors with encoders, both axes
Slew speeds	Nine slew speeds
Tracking rates	Sidereal, solar and lunar
Tracking modes	EQ North and EQ South
Alignment procedures	AutoAlign, 2-star align, quick align, 1-star align, last alignment, solar system align
Software precision	24 bit, 0.08 calculation
Computer hand control	Double line, 16 character Liquid Crystal Display; 19 fiber optic backlit LED buttons
Database	40,000+ objects, 100 user defined programmable objects. Enhanced information on over 200 objects
Power requirements	12 VDC 3.2 Amps
GPS	Optional SkySync GPS Accessory

CGE Pro Mount

Tripod and pier weight	52 lbs (24 kg)
EQ mount weight	75 lbs (34 kg)
Counterweight bar	5 lbs (2.27 kg)
Weight of counterweights	1 × 22 lbs
Weight (lbs)	154 lbs (70 kg)
Payload capacity	90 lbs (41 kg)

(continued)

(continued)

Motor drive	DC servo motors with encoders, both axes
Slew speeds	Nine slew speeds 5.5 deg/s, 2 deg/s, 0.5 deg/s, 64x, 16x, 8x, 4x, 1x, 0.5x
Tracking rates	Sidereal, solar and lunar
Tracking modes	EQ north and EQ south
Alignment procedures	2-Star align, solar system align, last alignment, quick align, 1-star align
Computer hand control	Double line, 16 character liquid crystal display; 19 backlit LED buttons
Database	40,000+ objects, 100 user defined programmable objects. Enhanced information on over 200 objects
Power requirements	12 VDC 3.5A
Internal clock	Yes
GPS	Optional SkySync GPS accessory

Appendix I

Glossary

AltAz	Altitude and Azimuth mount. Features the intuitive left–right and up–down movements of the telescope. Many GoTo telescopes operate in an alt-az configuration.
AFOV	Apparent Field of View. Usually applied to telescope eyepieces.
Android	A mobile operating system developed by Google for mobile devices such as tablets and smartphones.
Autoguider	This function processes the signal from a CCD camera installed on a guide scope, and it automatically guides the telescope and mount with high precision over an extended period. This enables long exposure photography and imaging of astronomical objects.
Backlash Compensation	Provides a reduced time lag at the point of revised motion where the mount drive gears briefly lose contact.
Dovetail	An almost universally accepted method of attaching telescope optical tubes to telescope mounts, originally introduced by Vixen, and now adopted by many manufacturer's including Celestron.

© Springer International Publishing Switzerland 2016
J.L. Chen, *The NexStar Evolution and SkyPortal User's Guide*,
The Patrick Moore Practical Astronomy Series,
DOI 10.1007/978-3-319-32539-2

Caldwell Catalog	A list of 109 bright deep sky objects not included in the Messier Catalog, compiled by Sir Patrick Caldwell-Moore. The list contains 35 galaxies, 9 nebulae, 13 planetary nebulae, 25 star clusters, 1 dark nebula, 18 globular clusters, 2 supernova remnants, and 6 star cluster/nebula combinations. The Caldwell Catalog includes bright deep-sky objects visible in the Southern Hemisphere that were not listed in the Messier Catalog.
Equatorial mount	Features the ability to track an astronomical object by countering the rotation of the Earth. The RA, or right ascension, axis is set parallel to the Earth's axis. The declination axis is the axis of rotation that is at right angles to the polar axis of an equatorial mounting and that permits pointing the telescope to celestial objects of different declinations. Declination is the measurement of an objects angular distance from the celestial equator.
FMC	Fully multi-coatings. Multi-layered anti-reflective coatings on all optical surfaces of a lens system. Applicable to eyepieces, refractors, and binoviewers.
FOV	Field-of-view. The true FOV is found by diving the AFOV of an eyepiece by the magnification that results from using the eyepiece.
GoTo Mount	Computerized telescope mounts that automatically point the telescope towards the requested object.
GPS	Global Positioning System. A system of 24 orbiting satellites (plus four spares) that use timing signals between them and to ground receivers to accurately, within 1 m, determine positioning on the Earth's surface. Also known as SATNAV.
HVAC	Heating, Ventilation, and Air Conditioning
IC Catalog	5386 deep sky objects cataloged by J.L.E. Dreyer as a supplement to the NGC catalog.
iOS	A mobile operating system developed by Apple for mobile devices such as tablets and smartphones.
Kings rate	The tracking rate developed to account for atmospheric refraction.

Meridian	An imaginary line drawn from due South directly overhead to due North.
Messier catalog	A list of 110 (actually 109) deep sky objects originally created by Charles Messier in the late 1700. It consists of 39 galaxies, 7 nebulae, 5 planetary nebulae, and 55 star clusters.
Moon or lunar rate	Sidereal rate plus compensation for the Moon's motion around the Earth.
NGC or New General Catalogue	A catalog of deep sky objects based on William Herschel's Catalog of Nebulae. The NGC catalog contains 7840 objects, and was created by J.L.E. Dreyer.
OTA	Optical tube assembly. Usually includes optics (lenses and/or mirrors), tube, mounting bracket(s), and focuser.
PEC	Periodic Error Correction, which compensates for slight manufacturing errors in the mechanical gear of equatorial mount drive systems that causes an irregular motion of the tracking gear. PEC enables smoother tracking, especially for astrophotography and astro-imaging.
SATNAV	see GPS
Sidereal rate	The standard tracking rate for compensating for the Earth's rotation. This is the rate the stars move across the sky.
SkyAlign	Celestron's proprietary GoTo mount alignment algorithm, requiring the user to align the telescope with three stars or planets. It is not necessary to know the names of the objects using SkyAlign.
StarBright XLT	Celestron's proprietary optical coating, StarBright XLT reflective and anti-reflective coatings are the purest available (exceeding 99.99%) and include aluminum (Al), hafnium oxide (HfO2), titanium dioxide (TiO2), silicon dioxide (SiO2), and magnesium fluoride (MgF2).
SCT	Schmidt-Cassegrain design telescope. Also known as a catadioptric design, meaning a telescope with a combination of lenses and mirrors.
SkyPortal app	Celestron's free application, available for download to iOS or Android mobile smart devices Powered by Sky Safari 4, SkyPortal allows the user to control the Evolution series telescopes, or through the use of the optional SkyPortal WiFi module attached to certain other Celestron

	telescopes or German mounts, to control the telescope and mount over a WiFi connection. The SkyPortal app allows the user to explore the Solar System, 120,000 stars, 220 star clusters, nebulae, galaxies, and dozens of asteroids, comets, and satellites—including the ISS.
Slew, or Slewing	The movement of a telescope on its mount using its drive motors.
Solar rate	The tracking rate required to maintain the accurate tracking of the Sun. This rate differs from the sidereal rate.
TOTOTA	The author's acronym for turning off, turning back on, and trying the alignment process again.
Wall wart	An AC-to-DC power supply that plugs into a wall electrical socket.
WiFi technology	Integrated wireless Internet allows connection to smartphone or tablet with the SkyPortal app to control Celestron's Wi-Fi enabled products.

Bibliography

Published Works

Bakich, Michael E., "Behind the Scenes at Celestron", Astronomy Magazine, December, 2013.
Celestron, *NexStar Evolution Series Telescopes Instruction Manual*, 2014.
Chen, James L., *The Vixen Star Book User Guide,* Springer, The Patrick Moore Practical Astronomy Series, 2015.
Harrington, Philip S., *Star Ware—The Amateur Astronomer's Guide to Choosing, Buying, and Using Telescopes and Accessories*, 4th ed., John Wiley and Son, Inc., 2007.
O'Meara, Stephen James, *The Messier Objects,* Cambridge University Press, 2006.
O'Meara, Stephen James, *The Caldwell Objects,* Cambridge University Press, 2002.

Internet Sources

www.celestron.com, accessed October 2015—March 2016.
www.apple.com, accessed October 2015—March 2016

© Springer International Publishing Switzerland 2016
J.L. Chen, *The NexStar Evolution and SkyPortal User's Guide*,
The Patrick Moore Practical Astronomy Series,
DOI 10.1007/978-3-319-32539-2

Index

© Springer International Publishing Switzerland 2016
J.L. Chen, *The NexStar Evolution and SkyPortal User's Guide*,
The Patrick Moore Practical Astronomy Series,
DOI 10.1007/978-3-319-32539-2